GAINING GROUND

GAINING GROUND
Japan's Strides in Science and Technology

GEORGE GAMOTA
WENDY FRIEMAN

BALLINGER PUBLISHING COMPANY
Cambridge, Massachusetts
A Subsidiary of Harper & Row, Publishers, Inc.

International Standard Book Number: 0-88730-308-0

Library of Congress Catalog Card Number: 88-18721

Printed in the United States of America

Library of Congress Cataloging-in-Publication Data

Gamota, George.
  Gaining ground : Japan's strides in science and technology /
George Gamota and Wendy Frieman.
    p. cm.
  Includes index.
  ISBN 0-88730-308-0
  1. Technology—Japan. 2. Research, Industrial—Japan.
I. Frieman, Wendy. II. Title.
T27.J3G36 1988
609.52—dc19   88-18721
                                                    CIP

CONTENTS

LIST OF FIGURES

LIST OF ABBREVIATIONS

AI	artificial intelligence
AIST	Agency for Industrial Science and Technology
AT&T	American Telephone & Telegraph
CAD	computer-aided design
CAM-I	Computer Aided Manufacturing-International
CCIS	Common Control Interoffice Signalling
CST	Council for Science and Technology
DARPA	Defense Advanced Research Projects Agency
DOC	Department of Commerce
DoD	Department of Defense
DRAM	dynamic random access memory
ECL	Electrical Communications Laboratory
EDS	Electronic Data Systems
EEC	European Economic Community
EPROM	erasable programmable read-only memory
ERA	engineering research association
ERATO	Exploratory Research in Advanced Technology Office
ESPRIT	European Strategic Programme on Research in Information Technology
ETL	Electrotechnical Laboratory
FCC	Federal Communications Commission
FET	field effect transistors

FGCS	Fifth Generation Computer System
FGHC	Flat Guarded Horn Clauses
FMS	flexible manufacturing system
FTU	flexible transfer units
GE	General Electric
GM	General Motors
HEMT	high electron mobility transistor
IBM	International Business Machines
IC	integrated circuit
ICOT	Institute for New Generation Computer Technology
INS	Information Network Service
ISO	Open Systems Interconnection Model
ITT	International Telephone & Telegraph
JAIF	Japanese Association of Industrial Fermentation
JICST	Japan Information Center of Science and Technology
JRDC	Japan Research and Development Corporation
JTECH	Japanese Technology Evaluation Program
KDD	Kokusai Telegraph and Telephone
KTS	key telephone system
LAN	local area network
LEC	liquid encapsulated Czochralski method
LED	light-emitting diode
LSI	large-scale integration
MAN	metropolitan area network
MAP	Manufacturing Automation Protocol
MBE	molecular beam epitaxy
MCC	Microelectronics and Computer Corporation
MCVD	modified chemical vapor deposition
MESC	Ministry of Education, Science, and Culture
MIT	Massachusetts Institute of Technology
MITI	Ministry of International Trade and Industry
MOCVD	metallic-organic chemical vapor deposition
Mombusho	Ministry of Education, Science, and Culture
MPT	Ministry of Posts and Telecommunication
NASA	National Aeronautics and Space Administration
NEC	Nippon Electric Company
NIH	National Institutes of Health
NSF	National Science Foundation
NSSP	National Super Speed Project
NTT	Nippon Telegraph & Telephone

OJRL	Optoelectronics Joint Research Laboratory
OMVPE	organometallic vapor phase epitaxy
OPEC	Organization of Petroleum Exporting Countries
PBX	private branch extension
PC	printed circuit
PCM	pulse code modulation
RAB	Research Association for Biotechnology
RRB	Radio Regulatory Bureau
SAIC	Science Applications International Corporation
STA	Science and Technology Agency
SCARA	Selective Compliance Assembly Robot Arm
SQUID	Superconducting Quantum Interference Effect Devices
SRAM	single random access memory
UHMW	ultrahigh molecular weight
VAD	vapor phase axial deposition
VAN	value-added network
VCR	videocassette recorder
VLSI	very large-scale integration
WANs	wide-area networks

ACKNOWLEDGMENTS

Let me begin by expressing my great thanks to the individuals and institutions who helped me write this book and made its publication possible.

First of all, I would like to express my great appreciation to Drs. Frank Huband and Duane Shelton, from the National Science Foundation (NSF), who inspired me to undertake the writing of this book and provided support throughout this period.

The book is based almost entirely on the results of a U.S. government contract to Science Applications International Corporation (SAIC) to study and evaluate Japanese technology in a number of technical areas. The program goes under the name JTECH and has been running since 1983. I am deeply indebted to the Department of Commerce for initiating the program and to its program manager, George Mu, for leading it in its first year. Supporters within DOC included Lionel Olmer, Bill Finan, Bo Denysyk, Tim Stone, and John Blehar. NSF took over the leadership role of JTECH with the support of the President's Office of Science and Technology Policy (OSTP). I would like to thank Drs. John McTague and Maurice Roesch from OSTP for helping out at a crucial moment in JTECH's life and having the vision of its importance to U.S. competitiveness. They were ahead of their time in recognizing that a thorough understanding of our competitors is key to U.S. technological leadership in the world. As program manager at NSF, Dr. George Hazelrigg provided badly needed help in steering the new JTECH program through its early beginning. Critical financial support from the people at the Defense Advanced Research Projects Agency (DARPA) enabled the program to be carried out successfully. Drs. Bob Cooper, James Tegnalia, Craig Fields, Joe Luquire, and John Meson are cited for appreciating the significance of the program and for their financial help.

As important as the support of the sponsoring agencies is the contribution of the more than forty-four JTECH panelists who contributed most of the factual information used in this book. They came from a wide range of institutions: leading universities, key national and private laboratories, and U.S. government organizations. They represent the best of U.S. technical expertise and have provided a valuable service for all of us. All panelists and chairmen are cited at the end of this volume. In particular, I would like to thank David Brandin, Laurie Miller, Professor Harry Wieder, Dr. James Economy, Professor George Turin, and Professor Dale Oxender, who helped by commenting on this book and reviewing their areas of expertise.

I would also like to thank SAIC for providing the means for carrying out this program. Without their patience, support—both financial and moral—it would have been impossible to achieve the success that has been accomplished. Encouragement from the top, Drs. Robert Beyster, Ed Frieman, and John Penhune, has been critical. Drs. Tony Armstrong and Marty Fricke provided the program management in the first couple of years. Their help is greatly appreciated. Last but not least, I would like to thank my colleague, Wendy Frieman, who took over the JTECH program at a very precarious point of its life and almost single-handedly shepherded it through some very difficult times and helped make the program what it is today.

With respect to his book, Michael Tyler helped summarize and tie the various JTECH reports together, and Wendy Frieman contributed greatly in integrating the socioeconomic aspects with the technical issues. As a frequent visitor to Japan and other Far Eastern countries, and fluent in their written and spoken languages, she was able to complement the "dry" technical findings and help interpret them in a setting that takes into account Japan's history, culture, and past and current infrastructure.

Finally, I would like to thank my family, particularly my wife Christina, who had to put up with many a missing weekend in order for me to finish this book.

George Gamota
Lexington, Massachusetts

1 INTRODUCTION

The United States awoke with a start to the challenge posed by Japanese science and technology in the early 1980s, only after Japanese products had systematically replaced their U.S. counterparts on the shelves of American retailers and the floors of American factories. Thus began a period of intense hand wringing on the part of U.S. government officials, corporate executives, and leading scientists about the possibility that they would live to see Japan become a technological as well as an economic threat to the United States.

Even a cursory look at Japan's science indicators, at best an imperfect expression of true capability, fueled this concern. In 1965 Japan's research and development budget was only 6 percent of the American figure for research and development spending. By 1982 the ratio of nonmilitary research expenditures to gross national product was 2.43 percent for Japan as opposed to only 2.01 percent for the United States. The number of patents granted to Japanese corporations tripled in twenty years, while the comparable U.S. figure held constant. Finally, Japanese technology exports grew from less than $1 million in 1960 to over $569 million in 1983 (Lynn, 1986: 296).

At the same time, awareness grew that the United States had, for the most part, simply failed over the preceding two decades to pay significant attention to the content of Japanese research, not believing it to be either a source of original thinking or a source of serious competition. What the United States didn't know could, in fact, hurt it. The fact that during those same two decades Japanese scientists had enjoyed virtually unlimited access to U.S. research and development engendered considerable resentment and intensified the fear of this unknown competitor. Never mind the fact that Japan's rise was apparent to many; few in prominent circles wanted to admit this threat.

Thus, Japan's sudden (to some) entry into the international scientific and technical community posed difficult questions for policymakers and for U.S. industry. What was the relationship between growing bilateral trade imbalances and Japanese national science and technology policies? If the results of Japanese commitments to technological growth were just beginning to appear, what did the future hold? Should the United States respond by competing head-on, denying Japanese access to U.S. research, demanding more access to Japanese research, or proposing to collaborate with Japanese researchers? Each of these questions raised a host of others, none of which could be adequately answered without more information about exactly how good Japanese research really was. Apocalyptic assessments about Japanese competition abounded in the trade press and the mass media. Much of the reporting was long on opinion and short on thoughtful analysis. To spend the 1980s overstating Japan's technological strength would be as great an error as it had been to spend the sixties and seventies ignoring it.

In this climate of escalating fear and accumulated neglect of Japan's technology base, the U.S. government launched several initiatives dealing with Japanese science. One such initiative was the Japanese Technology Evaluation Program (JTECH), sponsored by a number of U.S. government agencies and started by Science Applications International Corporation (SAIC) in 1983. Its goal was to provide objective evaluations of Japanese research and development in selected high-technology fields or disciplines.

THE JAPANESE TECHNOLOGY
EVALUATION PROGRAM

JTECH began as a research contract sponsored by the U.S. Department of Commerce. Other government agencies, including the National Science Foundation, also contributed funds to the JTECH effort; in 1985 NSF became the lead government support agency.

The JTECH evaluations are intended to provide a solid technical foundation upon which government officials, corporate planners, and scientists in academia and industry can make sound judgments about major policy issues. (The reports themselves concentrate on technical evaluations of Japanese literature rather than on policy questions.) JTECH panels have also been instrumental in fostering an interest in Japan in the scientific community. Several panels have held special sessions at association meetings to discuss their JTECH panel findings and have published articles based on JTECH in leading technical magazines. These activities are intended to encourage more scientists to recognize the importance of monitoring Japanese technical developments on a regular basis.

The JTECH report process takes approximately nine months. Because its success depends on the caliber of the scientists who participate in the program, JTECH selects six to ten nationally recognized experts for each panel. Although JTECH benefits greatly from involving professionals who have visited Japan or know the Japanese language, the primary criterion in recruiting panelists has always been technical expertise.

At the beginning of the study, these experts assemble in Washington to agree on the report structure and on individual chapter assignments. They each spend approximately four weeks over six calendar-months reviewing Japanese technical literature and evaluating the quality of Japanese research vis-à-vis that of the United States. SAIC supplies the panelists with a range of Japanese sources, including but not limited to Japanese articles from major technical journals, papers from meetings of Japanese technical societies, reports from Japanese government laboratories, trip reports of leading U.S. scientists, and Japanese patents. Some of these materials are available in English; some require translation. For the most part, JTECH has found that adequate technical material is obtainable if one knows where to look for it and how to exploit it.

Panelists are expected to compare Japanese research efforts to corresponding U.S. programs whenever possible and are required to provide references to support their assertions. JTECH also encourages the panel members to go beyond reporting ongoing research activities and discuss the significance, either scientific or commercial, of those activities. Finally, panelists are asked to discuss the direction in which the work is going and the rate of change. Each panel chairman is responsible for assembling a chart summarizing Japanese current capability and rate of change.

Whenever possible, JTECH panelists use visits to Japanese companies, universities, or research institutes to supplement a review of the printed literature. The panel chairman is responsible for integrating the report and preparing a summary assessment that captures the spirit as well as the content of the individual chapters. After six months, JTECH holds a workshop at which the panel presents its findings to representatives from the technical and policy communities. The workshop proceedings consist of a presentation by each panel member of one report chapter, followed by discussion sessions. After the workshop, panel members incorporate the suggestions made by the workshop attendees and submit their final drafts. SAIC then publishes the panelists' findings, together with the panel chairman's executive summary and assessment, as a final report. Once the report is published, JTECH makes every effort to distribute the report to the interested research and policy communities.

By 1987 JTECH had completed assessments in six technology areas and was preparing to undertake another three during 1988. This book summarizes what has been learned from the program to date. To be sure, this is a small data base, but it provides quite a view of some specific technological areas of major importance to the U.S. economy—present and future.

THE BOOK'S ORGANIZATION

The book is divided into three sections. Chapter 2 focuses on the Japanese research and development infrastructure and discusses briefly the differences between the U.S. and Japanese approaches to national science policy. Finally, Chapters 3 through 8 form the bulk of the study. They summarize the content of the six published JTECH reports on computer science, opto- and microelectronics (non-silicon-based), advanced polymers, mechatronics, telecommunications, and biotechnology. Each chapter discusses why the technology is important; describes the key research organizations in Japan and the most important funding mechanisms for that technology area; compares Japanese work with U.S. work in the field (including, where possible, prospects for the future); and summarizes the differences between the U.S. and Japanese approach to, and organization of, the field or discipline in question.

The summaries are based on the reports as they were originally written. As of October 1987, the panel chairmen of the six reports all felt that their overall conclusions were still valid and had not been superseded by events either in Japan or the United States. We were completing the seventh JTECH report, which covers advanced computing (the Fifth Generation Program and related efforts), just as this book was going to press. The executive summary for that report is included as the Appendix.

AN OVERVIEW

This book is only a first step toward understanding Japanese scientific and technical capabilities. The JTECH process is being constantly revised to correct inconsistencies and shortcomings. One problem has been that because the investigations focus primarily on the technical quality of Japanese research, the organizational, economic, or societal aspects of Japanese science are in many cases not systematically treated and are unevenly distributed among chapters. This makes it difficult and perhaps dangerous to draw conclusions on the broader characteristics of Japanese science based only on six imperfect studies. Nevertheless, several themes recur. One is the contrast between the U.S. emphasis on (and strength in) basic and undirected research and the Japanese emphasis on (and strength in) focused development work with specific commercial applications. Many of the disciplines examined, especially computer science, opto- and microelectronics, and biotechnology had their genesis in American university departments where there is an established basic research tradition. U.S. government funding for basic research, even Defense Department funding, which is normally thought to have a strong mission orientation, has in the past supported conceptual and theoretical inquiry.

For a nation perpetually under pressure to reduce, rather than expand, national budgets, the decision to invest in basic research cannot come easily. Although it is generally accepted that investments in basic research ultimately benefit the economy, it is virtually impossible to predict how long it will take for that benefit to materialize. From many studies, five to twenty to thirty years is often the amount of time between basic research results and commercialization of a product based on those results. In some areas, such as software research, the time could be shortened, but in others it is painfully long.

Perceptions of the practicality of research can be misleading, since most predictions of the future of technology become hazy if the future is more than a few years away. It may also be easier to predict the effects of some basic research activities than others. Thus in the Bell Telephone Laboratories in the 1940s, it may have been easier to predict that success in solid-state research could lead to a device replacing the vacuum tube, with vast commercial implications, than it was to predict that research on information theory would revolutionize the telecommunications field, with equally vast repercussions. Forty years later, it is still difficult to point definitively to many products that are direct descendants of information theory, while the descendants of the first transistor are legion. Yet it is impossible to imagine modern telecommunications without the influence of both. Likewise, it might be relatively easy to predict the effects of materials research (such as the work described in this book on III-V compounds) on telecommunications. The effects of mathematical research on large-scale systems or queuing theory are more difficult to project.

The United States, for better or worse, has supported basic research whose results are easy to predict, *as well as* basic research whose results are more difficult to envisage. Japan to a large extent, for better or worse, has supported neither. The Japanese have been able to gain enough understanding of the basic scientific principles to pursue advanced development and product engineering without actually engaging in basic or applied research themselves. Japanese government support for research has been limited to programs with specific milestones, targets, and often prototypes. This approach can be attributed to shrewd economic calculations, to national insecurity about competing in science, or to any number of other factors. However, it is undebatable that this approach has clear commercial benefits, and it is difficult to make a good case for changing a strategy that seems to work so well.

It is equally undebatable that the future Japanese (and American) achievements depend on continued basic research somewhere in the world. The performance of Japanese scientsists overseas and the basic research activities now beginning to emerge domestically suggest that Japan has the capability to be a player, if not a major player, in the scientific as well as the technical community.

A second and related theme is the continued Japanese practice of sending students and professionals overseas for study; reading, carefully screening, and sifting foreign technical literature; and looking abroad for improved products and processes as well as for potential competitors, even in fields where Japan enjoys an apparently comfortable lead. This behavior pattern has never existed in the United States for countries who do not pose a military threat. This difference is one factor responsible for the Japanese lead over the United States in advanced development and product engineering in virtually every field JTECH has analyzed. Two particularly good examples are advanced polymers and opto- and microelectronics.

A third feature of Japanese technology policy is the time frame used for national planning. Several chapters argue that the Japanese take a longer view than the United States in choosing national technology programs and setting a national technology agenda. In fact, the science and technology activities of each country documented in the reports tell a different story. The United States does take a long-term (at least twenty to thirty years) view when allocating money for basic research. The distinction to be drawn is between applied, non-military research programs in the United States, which may die after a few years if no commercial results are visible, and comparable efforts in Japan, whose funding (whether from government or industry, and usually from both) is much more stable over a ten-year period. The Japanese commitment to stay with Josephson technology, for example, contrasts with IBM's virtual abandonment of the field after only a few years. The recent discoveries of high-temperature superconductors have changed the picture drastically, forcing IBM to rethink its position and possibly rebuild its effort, while the Japanese employ a large cadre of talented scientists capable of exploiting these new findings and coming out with the first Japanese commercial product in the computer area.

All of the JTECH reports discuss the role of the Ministry of International Trade and Industry (MITI) in national technology planning, and all agree that the MITI programs have helped focus Japanese industry. In microelectronics and polymers national programs have resulted in important contributions. In no area, however, was a MITI program the single reason for Japanese success (or failure). Most reports pointed to the fact that MITI research funds account for only a small percentage of national research spending, and, whereas it may be a disproportionately important percentage, any number of fields have enjoyed considerable success with no help from MITI. Sensors, both biosensors and sensors for mechatronics systems, are a good example. The MITI programs are the result, rather than the cause, of industry consensus and commitment to a given area.

All of the JTECH reports referred to Japanese manufacturing excellence and attention to engineering detail, and several pointed to Japanese treatment of manufacturing as a serious discipline, worthy of research, as a key difference between the two countries. This is reflected in Japanese superiority in many areas of mechatronics and in electronic components. Although there has been much talk in the United States about the need to pay attention to manufacturing more generally, there is little evidence of improvement in this area. Even today, leading U.S. universities have not been able (or willing) to entice the best students to go into this area. The same is true of the faculty. If the university infrastructure doesn't consider the areas important, neither will the faculty nor the students. It is a vicious cycle that has yet to be broken.

The chapters also reveal that university communities play very different roles in the two countries. It is difficult to determine whether the academic community in Japan is more or less useful to Japanese science and technology than U.S. universities are to U.S. science. It is largely a matter of expectations. Nevertheless, the difference is worth noting. In Japan, university science departments exist primarily to grant bachelor's, master's and occasionally doctoral degrees to students who will be trained much more extensively by their companies after they graduate. Graduate student and faculty research activities tend to be more applied than U.S. university research. University departments play a role in industry through informal mechanisms and arrangements involving specific faculty members rather than entire departments or institutions. The contributions made through these mechanisms have, however, been significant, especially in specific areas such as software for mechatronics and radio propagation.

In the United States, by contrast, the science departments at leading universities occupy a central role in many fields and technology areas. Their function extends far beyond training to performing both basic and applied research, and their relationships with government and industry have become institutionalized.

Finally, universities, laboratories, and corporations in the United States respond to military as well as civilian funding sources. Some of the best U.S. work cited in this book in fields such as high tenacity/high modules polymers and sensors was initiated to support a military objective and resulted in few if any commercial benefits. Targeted research funding with identifiable and deliverable results is possible to a large extent in the United States only for military projects. The fact that Japan can use the same mission orientation for commercial rather than military projects is one more cause of technology and trade imbalances between the two countries.

The above are more hypotheses than conclusions, and are based on very incomplete information. Additional data and analysis are required to test them systematically. After years of neglect, JTECH's work represents only a first step toward understanding Japanese science and technology; much remains to be done.

REFERENCES

Lynn, Leonard. 1986. "Japanese Research and Industrial Policy." *Science*. July 18.

2 THE CONTEXT FOR JAPANESE SCIENCE

JTECH has always focused on the technical quality of Japanese research; the conduct and management of science in Japan have been
at most a secondary concern. This chapter supplements the JTECH
report summaries by sketching out the policy and institutional context for Japanese science and technology achievements. A description of the major science policy-making organizations is followed by
a brief discussion of how these policies are implemented. The final
section of this chapter summarizes the major differences between
U.S. and Japanese approaches to science and technology.

KEY JAPANESE ORGANIZATIONS

Several Japanese government organizations have the responsibility
of formulating and carrying out science and technology policy.
Some have redundant functions, which leads to competition and
bureaucratic infighting. This section briefly describes the key organizations and how they cooperate and compete with one another.

The Science and Technology Agency

The Science and Technology Agency (STA), charged with promoting
science and technology that contributes to the national economy,
spends roughly one-quarter of the national science budget. The STA

is part of the Prime Minister's office. In order to promote research that will benefit the economy, the Agency coordinates research activities of the industrial ministries and funds projects that are vital to the nation but too large or risky for individual firms to undertake (Anderson, 1985). The STA spends most of its money administering public agencies engaged in development of nuclear energy, ocean exploration, and space technology; and operating six national laboratories.

The STA also administers two public agencies that are now the focus of increasing foreign attention by the early 1980s. The Japan Information Center of Science and Technology (JICST) maintains a large data base of Japanese and foreign scientific literature that is available to U.S. scientists through the National Technical Information Service.

The Japan Research and Development Corporation (JRDC) was established in 1961 in an effort to commercialize government-sponsored basic research. Companies who foresee long-term applications of government-sponsored research compete for funding from the JRDC. The loan must be repaid if the firm benefits commercially from the project, although the company does retain exclusive rights to products and processes developed with JRDC money for two or three years.

The JRDC also administers the Exploratory Research in Advanced Technology Office (ERATO), aimed at stimulating creativity in Japanese science and at fostering a more widespread interest in basic research. ERATO funds small, interdisciplinary teams of researchers for several years to work on highly innovative projects. The ERATO teams are encouraged to recruit scientists under the age of thirty-five as well as foreign scientists.

The Ministry of Education, Science, and Culture (Mombusho)

The Ministry of Education, Science, and Culture (Mombusho) accounts for roughly half of national spending on science and technology; of this, about two-thirds is allocated for research and teaching at twenty-one national universities. The ministry is responsible for funding the primary, secondary, and university education system and for formulating education policies to meet the manpower requirements of Japanese science and industry. Japan's major corporations recruit primarily from the University of Tokyo and a handful of other prestigious national universities; 80 percent of government professionals are hired from the University of Tokyo. Ninety percent of Japanese children remain in school until the age of eighteen, and more than 30 percent of Japan's high school graduates attend colleges or universities. Illiteracy in Japan is less than 7 percent (Lynn, 1986). Mombusho also oversees ten national research institutes, six of which perform basic research.

Based on long-term projections by the Science Council of Japan, Mombusho initiated a rapid expansion of science education at the university level between the mid-1960s and the early 1980s. University education is heavily weighted towards applied courses, often at the expense of basic or theoretical science, and Japanese universities now award at least as many bachelor degrees in engineering as American universities.

Mombusho is the one government organization that funds basic as well as applied research. It uses three funding channels to support science at national universities. First, it provides general-purpose research funds to each chair or division of a university department. Second, it allocates special research funds for construction of new plants and purchase of new equipment. Finally, it administers a grant-in-aid program for researchers or groups of researchers whose applications are approved by the Mombusho Science Council.

Mombusho's ten research institutes are administered under the National Research Institutes for Joint Use by Universities program. This was established in 1973 to provide research facilities run by a board with national representation and available to a wide user base. These institutes have some features of university-based research centers and some features of government research institutes.

Japanese universities operate on a chair or *koza* system. A *koza* consists of a full professor, several associate professors, two assistants, and two technical staff members, all of whom have civil service status. The *koza* is funded as a unit to provide teaching and perform research in a particular area; it operates independently of other *kozas* and other nonuniversity research entities. Although *kozas* are free to set their own agendas, they are so small that their budget often precludes buying expensive equipment and instrumentation.

Recently Mombusho has tried to foster greater interdisciplinary cooperation within the large universities and more extensive collaboration between industry and government. Numerous disincentives to such cooperation exist (Samuels, 1987). If a *koza* accepts corporate funding for scientific research, for example, its budget from Mombusho is automatically reduced by that amount. However, university professors can and do accept the donation of scientific equipment as well as consulting fees (not legally permitted according to existing civil service codes) as payment for research services. The extent or importance of university professors in Japanese industrial research is still extremely difficult to gauge.

Ministry of International Trade and Industry

MITI spends approximately one-eighth of the national science and technology budget (Lynn, 1986). Although most of the other ministries have an interest in science and set aside funds for research and development programs, MITI is extremely powerful in setting the course of industrial research in Japan. When the technology gap between Japan andthe West was still quite large, MITI's role in science and technology was to identify, acquire, and disseminate new foreign technology. Now that the gap has narrowed in many fields, MITI's role, carried out for the most part by its central bureaus and by its Agency for Industrial Science and Technology (AIST), is to coordinate research done by other ministries, promote and facilitate industrial research by the private sector, and operate national laboratories.

MITI has used a variety of mechanisms to promote industrial research. The one most widely publicized outside Japan is the national research project, in which a number of firms in a given technology area join together, sometimes at a jointly operated research laboratory, to work on research problems early in the development of a new industry. These projects, many of which are sponsored by AIST, can take two forms. One is an engineering research association (ERA), which is awarded special tax privileges and conditional loans. The very large-scale integration (VLSI) program was one such arrangement. The second form is a public corporation (such as the Institute for Fifth Generation Computing, or ICOT), which is also entitled to certain privileges. MITI typically initiates national projects in technology areas in which the near-term research costs are too high and the prospects too risky for any one company to be able to make significant progress, yet whose long-term market potential is widely believed to be very strong. Funding for national projects comes either in the form of grants or success-conditional loans, which need to be repaid only if the companies make a profit on the resulting product (Samuels, 1987).

Japanese corporations often assign their top people to participate in MITI projects. Recruiting professionals directly from the participating companies is one mechanism for transferring what is learned at a national project back to the company. Participating scientists generally return to their corporation after the project is over to work on product development. Because MITI provides strong coordination, it is often possible to avoid wasteful duplication of effort and promote speedy diffusion of technical results through symposia, conferences, and nondiscriminatory access to patents. National projects are a mechanism for bringing together science professionals in a system where there is lifetime employment at most large companies, relatively little labor mobility in the technical work force, and comparatively little informal interaction across company or institute lines.

The concept of a national project in which several competing high technology firms collaborate on technical problems has received a great deal of attention in the West; it was the inspiration that led to the creation of the Microelectronics and Computer Technology Corporation (MCC) in Austin, Texas. Several little-known features of Japan's national projects are worth mentioning.

First, not all national projects involve the establishment of joint laboratories. Only three of the eighty-two ERAs even sponsored joint laboratories, and many projects, such as the AIST program on Basic Technologies for Next Generation Industry, simply fund existing institutes and corporations. Second, even where joint laboratories exist, the lines between shared generic technology research and closely guarded proprietary research are carefully drawn. Representatives from the participating corporations spend a great deal of time at the outset of a project deciding which problems will be addressed jointly and which will be left for individual companies to pursue. Third, even when working on those problems that are agreed to be generic and nonproprietary, participating companies often work on different pieces of or approaches to the problem (Tucker, 1985). Seldom if ever do competitors actually work side by side on exactly the same problem for an extended period of time. Fourth, not all companies accept the invitation to participate in MITI-sponsored projects or joint laboratories. Corporations are sometimes wary of participating in a joint research effort, especially if the project is in an area where the company has a comparative advantage due to unique technological strength. Finally, MITI decides to pursue a national project only after lengthy and careful consultation with industry leaders, usually through meetings of MITI's Industrial Structure Council. This forum is a mechanism for industry representatives to express their views about the value of technological alternatives and approaches, and it helps build the consensus required for successful projects. MITI bureaucrats tend to have a background in law or economics, not science and technology, so they listen carefully to technical experts before allocating resources.

All government-commissioned research is the property of the Japanese government. However, licenses are available to firms on a nondiscriminatory basis. If a government loan to a private laboratory results in a patentable discovery, the private organization owns the property rights and must repay the loan. In joint research, decisions as to which company owns the property rights are made case by case (Samuels, 1987).

In addition to sponsoring national projects, MITI offers other incentives for both single-company and collaborative research and development. It is usually able to exempt companies from antitrust restrictions on collaborative research. Antitrust restrictions, in general, tend to be more casually enforced in Japan than in the United States. Attractive loans are available to companies who join together on their own to perform collaborative research even in areas where there is no national project. These firms can depreciate capital equipment used in collaborative research by a full 100 percent in the first year of the project. Research consortia can own their own equipment valued nominally at 1 yen and can enjoy a drastic reduction in fixed property taxes after the first three years of the project.

MITI also lobbies heavily for incentives to promote single-company proprietary research. The Japanese tax code has at least nineteen different provisions intended to stimulate innovation and investment in research and development. A research credit is available to cover expenses incurred between 1985 and 1988 for research and testing into the improvement, study, or discovery of manufacturing technologies. The most recent provision is a tax credit equal to 7 percent of the acquisition cost of assets used in the development of 126 specified technologies ranging from cellular engineering to semiconductor pattern etching (Stern, 1987). The degree to which these incentives give Japanese corporations an edge over their U.S. competitors is still a subject of considerable debate. Many structural asymmetries exist between the U.S. and the Japanese tax codes, making one-to-one comparisons of specific provisions extremely difficult (Saxonhouse, 1986).

The task of operating Japan's national laboratories falls to AIST. These sixteen laboratories employ close to 3000 scientific professionals and account for nearly one-fourth of the budget for all government research institutions. Recently MITI introduced two programs designed to enhance cooperation between AIST labs and private sector firms. The first is a fund to provide local firms access to AIST equipment and to introduce commercial themes to the government laboratories. The second is a program under which interns from the private sector participate in AIST laboratory projects for one to three years.

The Science Council of Japan

The Science Council is an independent body of Japanese scientists who advise the Prime Minister on all aspects of science and technology policy. The members are elected by the scientific community and serve a three-year term. Although the Prime Minister must review and accept proposals of the Science Council, in practice he can postpone action on its recommendations indefinitely. The Science Council has seen its influence decline since the establishment of the Mombusho Science Council in 1962. The Science Council has taken political positions considerably to the left of the government, especially on issues such as nuclear power, and has been accused of using the organization to promote an ideological rather than a scientific agenda. This has diminished its influence even further (Anderson, 1985).

The Council for Science and Technology

The Council for Science and Technology (CST) is a supraministerial organization responsible for coordinating across the vertical lines of the Japanese bureaucracy. The Council is essentially an arm of the

Prime Minister's office and is chaired by the Prime Minister. Some of its members serve ex officio, including the Minister of Finance, the Director-General of the STA, and the president of the Science Council. In addition, the Prime Minister appoints five members who are nationally recognized scientists. The CST is charged with formulating the broad policy framework for science and technology in Japan and with defining long range goals for Japanese scientific research. The general objectives of Japanese science have changed little over the past fifteen to twenty years. They are developing domestic sources of energy, reducing dependence on foreign raw materials, stimulating research in high technology, and improving innovative capability through better coordination of academic, industrial, and government research.

Relationships among Government Bodies

The preceding description suggests that government science organizations have overlapping responsibilities. A more complete picture of the Japanese science system would only reinforce that view, as many other agencies, including the Ministry of Posts and Telecommunication (MPT), the Ministry of Transport, and the Ministry of Health and Welfare, also have research responsibilities. The competition to be designated the lead agency for a national project is extremely keen. Government agencies even stake bureaucratic claims by calling press conferences to announce projects before they have been approved by the Ministry of Finance.

Japanese Corporations

Japan's major industrial corporations are critical to national science and technology planning for two reasons. First, corporate spending on research and development accounts for more than three-quarters of total Japanese spending on science and technology. Companies who are participating in national projects expend their own resources on research that strengthens or complements their participation in these larger projects before, during, and after the life of the project. Moreover, as mentioned earlier, corporate representatives meet regularly with MITI councils to report on technology developments and project new markets. Thus, both internally and through the MITI councils, private corporations make critical investment decisions that will shape the future of Japanese technology.

Second, corporations provide extensive technical training for their employees, much more so than is common in the United States or Western Europe. Graduate students often accept positions in corporations and receive their degrees later, after company training, on the basis of a thesis. Such degrees carry the same weight as degrees granted to resident students. The responsibility for training also gives companies a major voice in the scientific future of the country. It is important to remember, however, that only the largest Japanese companies invest heavily in research and training. The tiers of small and medium-sized subcontractors rely on their customers for research and development.

Large Japanese corporations typically conduct research at two levels within the company. The first is a central research laboratory where scientists work on basic or generic problems. The second level is the divisional laboratory, where researchers work on problems related to product development and manufacturing. It is not uncommon for scientists from the corporate level laboratory to see a product through the entire development cycle, transferring first to a divisional laboratory and then into marketing or sales. Such transfers are a source of prestige and often entail an increase in salary (Westney and Sakakibara, 1985).

COMPARING U.S. AND JAPANESE SYSTEMS

This overview of the organizations involved in Japanese research and development suggests differences between U.S. and Japanese approaches to science and technology. Distinctive features of the Japanese system include the important role played by corporations in training technical personnel; the continuing process of consultation between MITI and industry in formulating national research priorities; the serious commitment by companies to research and development; the limited mobility of the Japanese technical work force; and the commitment to the normalization of knowledge facilitated by collaborative research. However, the two countries are also different in other ways that directly influence the formation and implementation of science and technology policy. Differences in science reflect distinctly different historical factors, economic requirements, and political orientations of the two countries. Three of these are worth introducing here.

The late exposure of Japan to Western science and culture resulted in a national commitment to "catching up" that has not existed in the United States for decades. This orientation has greatly influenced Japanese achievements in science and technology, in part because it required efficient use of foreign knowledge. The technology gap between Japan and the United States was for many years a powerful incentive to scrutinize foreign technology, acquire it at the most advantageous prices, and adapt it to fit the needs of the domestic and international markets. This behavior continues even as the gap is closing in many areas. Late development has made monitoring foreign technology a legitimate—if not critical—function of Japanese government and business. Considerable resources are devoted to overseas fact-finding missions, data base development, and systematic analysis of foreign scientific, technical, and business literature. Although Japan is now a net exporter of technology, the Japanese still acquire foreign products and processes that complement indigenous capability. In sum, Japan has a strong predisposition to study, acquire, and learn from foreign technology. No such sentiment exists in the United States, despite increasing evidence that a complacent attitude has had serious economic consequences.

A second difference, closely related to a reliance on foreign technology, is the emphasis on applied rather than basic research. The United States has a strong basic research tradition, fostered in part by defense programs which have demanded extensive theoretical work and in part by the priority placed on maintaining and equipping high-quality technical universities. Japan has learned from studying the basic research of others and has made few (some would say no) theoretical breakthroughs. The Japanese fear that they will begin to lose the access they have enjoyed to foreign, especially U.S., science. Ultimately, they believe, economic growth will have to emerge from purely indigenous research activity. For this reason, the Japanese government is attempting to promote creativity in science and basic research. Nevertheless, the overwhelmingly applied orientation of the scientific world in Japan cannot change overnight, and it is likely that much of what is considered basic research in Japan would be classified as development or applied work in the United States. A clear illustration of this difference in perception is the

ERATO program described above which, although it is described as basic research, has generated over 100 patent applications (Samuels, 1987).

An equally important difference is the absence in Japan of a sizable defense budget. Weapons procurement by the U.S. military has been a powerful force shaping the development of science and technology in the United States for the past thirty-five years. In Japan, since the reorganization of Nippon Telegraph and Telephone (NTT), there have been no large government hardware procurements which dictate a scientific research agenda. Instead the Japanese government has developed, in consultation with industry, national research priorities whose applications are purely commercial. Likewise, in the absence of a compelling domestic market for military hardware, Japanese corporations invest their own resources in science and technology that has civilian rather than military applications.

To what degree are these asymmetries responsible for differences in scope and quality between Japanese and U.S. research? The answer varies with each discipline or technology area and will, in the final analysis, be somewhat subjective. The previous chapter offered some tentative conclusions based on the JTECH reports prepared to date.

REFERENCES

Anderson, A. 1985. *Science and Technology in Japan.* London: Longman.

Lynn, Leonard. 1986. "Japanese Research and Industrial Policy." *Science* 233 (July 18): 296–300.

Samuels, R. 1987. "Research Collaboration in Japan." MIT Japan Science and Technology Program WP 87–02.

Saxonhouse, G. 1986. "Why Japan Is Winning." *Issues in Science and Technology* (Spring): 72–79.

Stern, J.P. 1987. "Japan's R&D Tax Credit System." ACCJ Journal. April. pp. 20–23.

Tucker, J. 1985. "Managing the Industrial Miracle." *High Technology* (August): 22–30.

Westney, E., and K. Sakakibara. 1985. "Designing the Designers." *Technology Review* (April): 26–68.

3 COMPUTER SCIENCE

Japanese entry into the computer field has profound commercial implications. The market for information systems is now as large outside the United States as it is within the country and, until recently, the United States has been the prime beneficiary of this worldwide business and has dominated all aspects of the computer industry. This dominance has been exemplified by International Business Machines (IBM), which controls about one-third of the worldwide market for computer and related products. IBM is the market share leader in every country in which it competes—except Japan.

In recent years, a number of Japanese companies, led by Fujitsu, Hitachi, and Nippon Electric Company (NEC), have made significant inroads into U.S. and European markets. Fujitsu owns nearly half of the U.S.-based Amdahl Corporation and is its primary supplier of computer equipment. Similarly, Hitachi is the source of most products sold by National Semiconductor's computer subsidiary. NEC supplies most mainframe computer products sold in the United States by Honeywell and in Europe by Groupe Bull.

These Japanese companies also control their home market. Both Fujitsu and NEC gave larger market shares than IBM, and Hitachi, Oki, and Toshiba are close behind. In the personal computer arena, the Japanese have been less successful commercially, although here, too, market penetration is beginning. The personal computers sold by Unisys and Leading Edge are made by Mitsubishi, and several low-cost Japanese (and Korean) models are currently eroding IBM's market leadership in the United States.

This chapter evaluates Japanese research in computer science as one of the critical factors in the future of the industry in both countries.

ORGANIZATIONS INVOLVED IN JAPANESE COMPUTER SCIENCE RESEARCH

MITI Support and Private Sector Computer-Related Research

Computer science research in Japan has benefited both directly and indirectly from the electronics research funded by MITI and the leading electronics firms. The VLSI project lasted from 1976 to 1980 and involved seven major computer manufacturers; the government-run Electrotechnical Laboratory (ETL); and Nippon Telegraph and Telephone's Electrical Communications Laboratory (ECL). As a result of this project, Japanese companies were able to take 70 percent of the market in 64K dynamic random access memories (RAMs), the building block of computer memories. Two follow-on projects include the High-Speed Computing System Project and the Fifth Generation Computer Project (ICOT). Many of these cooperative efforts require a prototype as one of the project milestones, a requirement that is rarely stipulated in U.S. government funding of computer science research. The High-Speed Computing project is discussed in greater detail below; a JTECH summary assessment of ICOT can be found in the Appendix.

Japanese companies conduct extensive in-house computer research and development programs to supplement joint and MITI-sponsored programs, spending between two and three dollars for every dollar spent by MITI. The major Japanese computer companies performing important research are Fujitsu, Hitachi, NEC, Oki, Toshiba, Mitsubishi, and Matshushita.

University Research in Computer Science

Major universities involved in Japanese computer science research include Keio, Kyoto, Tokyo, Toyohashi, and Tsukuba. A Science

City was recently built adjacent to the University of Tsukuba. Funding for university-based research comes from two sources: the Ministry of Education and from AIST. The quality of the equipment found at the department level is inferior to that in U.S. university departments. However, the central Japanese university computer centers are better funded and equipped than their U.S. counterparts. Many of them have top-of-the-line products and prototypes from Japanese manufacturers, which benefit both the university (through better technology) and the computer firms (through visibility and feedback from users before final production). For the most part, the universities operate systems compatible with IBM hardware and software, although some are beginning to switch to the Digital Equipment Corporation's Unix environment currently favored by U.S. universities.

The single striking contrast between Japanese and American universities is in graduate training and research programs. In the United States, much of the important basic research in computer science is conducted in universities. In Japan, basic research is done in industry, and the graduate student body in computer science is relatively small. Graduates are hired and then trained to do research specific to the needs and objectives of the company. This arrangement is made possible by Japan's tradition of lifetime employment in large companies.

Professors are a central element in transferring computer technology from universities into industry. They serve on national committees, on the ICOT board, and in other positions that affect the future of the industry. Although technically they are not permitted to accept consulting fees, they provide advisory services to computer firms in exchange for honoraria or laboratory equipment.

EVALUATION OF JAPANESE COMPUTER SCIENCE RESEARCH

The next sections of this chapter evaluate the progress Japanese institutions have made in four important segments of computer science: processor architecture, software, artificial intelligence, and communications.

Processor Architecture

The hardware of a sophisticated information processing system is the architectural design of the central processing unit. This design includes the basic arrangement of semiconductor-based integrated circuits for computational tasks; the coordination of the central computational elements with information storage, retrieval, and transmission elements; and the physical requirements regarding temperature control, space allocation, and cost.

Basic Research. Japan has clearly made significant technological and commercial breakthroughs in product engineering of processor architectures. Yet the Japanese effort in basic research lags behind U.S. work and is falling farther behind. It seems likely that Japanese researchers will continue to rely on Western basic research in their prototype development and laboratory experimentation.

Advanced Development. Despite a dearth of basic research, Japan is on a par with the United States in the advanced development of prototype architectures, successive refinement of those architectures through laboratory experimentation, and high-quality, low-cost manufacturing. Japan is likely to focus its resources on these areas and is projected to improve its position relative to the United States in the near future.

Japan's interest in advanced development of processor architectures spans virtually all types of architectures, although the application of these architectures in actual products shows a bias toward certain product categories. Four distinct types of processor architectures can be identified: sequential, parallel, distributed, and specialized. In sequential architecture a single, central processor performs one operation upon one piece of data at a time. In parallel architecture a central processor performs the same operation upon several pieces of data at the same time; this provides faster processing but limits the types of computations that can be made. Distributed architecture involves several processors operating in conjunction with a network, generally performing different operations upon different data at the same time. No single element controls the entire network. Specialized architectures include architectures designed for artificial intelligence, signal processing, and similar applications.

The most prominent advanced development has been aimed at sequential and parallel architectures. In sequential architectures, for example, Japan has shown interest in both microprocessor-based and mainframe designs. Japanese advanced development in this area is gaining ground, and product engineering far exceeds U.S. capabilities. Most prominent in this area are Fujitsu's and Hitachi's successful implementations of IBM's mainframe architecture. These implementations do not merely mimic IBM hardware; they also provide additional capabilities for enhanced functionality. NEC has also developed highly advanced mainframe and microprocessor architectures that have met commercial acceptance.

In the realm of parallel architectures, a wide variety of research has been reported. Japan has generally lagged behind the United States in all areas of parallel architecture research, but it is improving significantly in advanced development. This includes development of models for parallel execution of logic programs, general scientific applications, sparse matrix problems, linear equation solving, and distributed data base applications. The University of Tokyo, for example, has developed a "parallel inference engine" for logic programming, and the University of Tsukuba has reported on research involving its PAX-128 highly parallel processor array for scientific applications.

Japan conducts important R&D in distributed and specialized architectures. Oki Electric has published work on a distributed data-driven processor, and research at the University of Toyohashi is exploring regular tree expressions and their mapping to functional network architectures. NEC has led the Japanese effort in the development of special-purpose architectures for artificial intelligence applications.

Product Engineering. In some areas, Japan's product engineering work equals or exceeds that of the United States—but it tends to be somewhat narrowly focused. Japan has no product engineering effort in the area of parallel architectures that matches the U.S. efforts of firms such as Concurrent Computer Corporation, Convex Systems, or Encore Computer. Whereas workstations constitute an entire industry in the United States, ICOT's Sequential Inference Machine prototype is one of Japan's few attempts to develop a professional workstation.

Japan's product engineering work is highly competitive in some areas. The Hitachi and Fujitsu mainframes built around their IBM-compatible sequential architectures have begun to win market share away from IBM. The Fujitsu machine, in particular, has important capabilities that customers have found highly attractive. One is that customers can partition Fujitsu mainframes so that a single machine can run two wholly independent operating systems concurrently, as if it were two machines; IBM's mainframes lack this capability. Fujitsu's capability is especially attractive to customers who are currently upgrading their operating system from MVS/SP (which uses 24-bit addressing) to MVS/XA (which uses 31-bit addressing). By enabling its mainframes to run both XA and SP simultaneously, Fujitsu lets its customers effect an orderly migration to XA over an extended period of time without interrupting important programs running under MVS/SP. IBM's lack of a similar partitioning capability may be due not to a technology problem, but to a marketing imperative; if no such capability is available, customers will need to purchase a second mainframe in order to migrate comfortably to MVS/XA.

Supercomputers. Perhaps the most significant Japanese efforts, however, are in the supercomputer arena. Developed initially by U.S. firms, supercomputers are built upon a highly specialized vector architecture, taking the concept of parallel processing as far as is practical. These computers can operate at speeds orders of magnitude faster than traditional mainframes built around sequential architectures, but their requirements for vectorized data have so far limited their useful application to problems such as weather forecasting, seismology, and ballistics. The most prominent users of supercomputers to date have been the U.S. Department of Defense, national laboratories such as those at Los Alamos and Livermore, and major U.S. universities known for pioneering basic research. The need for continuing technological advancement in this area is therefore both apparent and pressing.

In recent years, however, the Japanese have suddenly appeared to challenge U.S. supremacy in supercomputer research. Japanese basic research in supercomputer hardware is now considered to be equal to that of the United States. In advanced development and product engineering, the Japanese are slightly behind, but improvement is rapid in both areas.

Supercomputer systems consist of system hardware, system software, applications software, and the dynamic interaction among these elements. In recent years, substantial advances have been made in each of these elements. Taken together, these improvements have resulted in an increase in performance, at constant cost, of six orders of magnitude in the past four decades.

The research underlying these advances has taken many forms. In hardware, new designs have improved the speed of component elements, and have increased the systems' concurrency; that is, supercomputers complete each operation faster and can perform more operations simultaneously. In applications software, improvement has come through the reduction of the number of steps needed to achieve a desired result and through the more effective use of hardware capabilities. Systems software improvements have been noticeably lacking; a consensus exists that FORTRAN, the most commonly used language, has yet to be optimized effectively anywhere in the world for use with supercomputer hardware and applications programs.

Two U.S. companies are responsible for U.S. advances in supercomputer research: Cray Research and Control Data (along with its ETA subsidiary). Through the mid-1980s, they had been entirely responsible for all supercomputer R&D in the world, and they held a seemingly invincible lead in all areas of research. By the end of 1983, however, Fujitsu, Hitachi, and NEC had all shown a clear commitment to entering and leading the field. By summer of 1986, Fujitsu was shipping commercial supercomputers to U.S. customers and NEC had sold its first supercomputer to the Houston Area Research Consortium.

The Fujitsu supercomputer is especially noteworthy, since it is the first supercomputer in the world that can run both specialized applications and programs written for general-purpose IBM mainframes (which have a sequential architecture). Analysts estimate that perhaps 20 percent of IBM mainframes are used for scientific applications, although IBM has never manufactured or sold a supercomputer. It does sell a vector processing attachment to its high-end mainframes. With the capability of processing programs written for IBM machines, the Fujitsu product could replace many IBM mainframes and preempt Cray and Control Data from a potential new market.

Japan's High-Speed Computing System project was announced in early 1981. This project has number of goals involving the development of supercomputer technology. It is jointly funded by AIST and industry and administered by a research association. This program hopes to have a prototype machine in 1989 and to develop technology with potential applications beyond supercomputers.

Comparable U.S. and European Efforts. Recognizing the strategic importance of research in parallel architectures, both American and European industries have every intention of meeting the Japanese challenge. In the United States, Cray and Control Data have stepped up their research efforts. Cray is even working on two entirely separate supercomputer development projects, one using conventional silicon-based technology and the other built on gallium arsenide. Research is also being conducted by the Microelectronics and Computer Technology Corporation (MCC), a collaborative research venture. MCC's four major research programs address advanced computer architecture, software technology, computer-aided design and manufacturing (CAD and CAM), and system packaging. The venture is wholly funded through the approximately twenty participating companies.

The United States has also responded through the Strategic Computing Program, an initiative of the Department of Defense Advanced Research Projects Agency (DARPA). Whereas the Japanese and American industrial efforts are developing supercomputers for a broad array of applications, DARPA is strictly interested in military computing. Starting with a set of perceived requirements for military computing in 1990, the program has identified eight targeted areas for advanced development. These include prototyping, parallel architectures, microsystem design methods, and microelectronic fabrication technology. Research in these areas is expected to be carried out by industry and university labs. The program had a $50 million appropriation in 1984.

The European Strategic Programme on Research in Information Technology (ESPRIT) is sponsored by the European Economic Community (EEC). Similar to DARPA's Strategic Computing Program, it hopes to stimulate industry and university research in advanced microelectronics, software, advanced information processing (akin to the Japanese fifth generation effort), and other areas. Each project must involve at least two EEC countries and be funded less than 50 percent by the EEC. Total research funding in the pilot phase, from 1983 to 1984, was about $10 million, but the main program may cost as much as $300 million annually.

A second European effort is the Alvey Programme for Advanced Information Technology, a consortium of British firms and universities managed by the United Kingdom's Department of Trade and Industry. The program is largely commercial in nature, designed to improve the British competitive position in software engineering, computer-aided design and manufacturing, man-machine interfaces, and knowledge-based systems; the last two areas overlap considerably with Japanese research in parallel processors and fifth generation computing. Some $480 million has been projected in funding over the program's first five years.

Software

Software research can be segmented into five major areas: software engineering, operating systems, applications packages, programming languages, and data base technology. Software engineering is concerned with the construction of multiperson, multiversion, multiyear programs. Operating systems are the most basic software instructions governing the computer system's course of operation. Applications software packages enable users to direct their computers to perform specific tasks or to manage specific functions, such as payroll and financial management. Programming languages are the vehicles through which applications packages are created by programmers. Data base technology allows users of large systems to organize and retrieve massive amounts of information in a coherent manner.

Basic Research and Advanced Development.   Japan conducts virtually no discernible basic research in any of these areas, nor is any such research likely to be instigated in the near future. The only significant exception is in the area of software engineering, where a number of "software factories" have done an excellent job of translating research results into practice.

Product Engineering.   Current product engineering in Japan represents a mixed bag, but the overall level of sophistication and interest is rather low. There seems to be no coordinated plan of research in operating systems, programming languages, or applications software in Japan. The lack of applications software can be tied to language barriers, since such software represents the direct English-language interface between the computer system and the American user. The Japanese aversion to research in programming languages and operating systems, however, cannot be tied to a language barrier. Rather, if it is taken in the context of Hitachi's and Fujitsu's strategic decisions to market IBM-compatible computers, this lack of effort appears to follow market imperatives. Software developed for IBM mainframes will run unmodified on Hitachi and Fujitsu products, thereby eliminating any need these companies would have for developing their own systems software.

NEC, by comparison, has contributed major elements of the ACOS operating system used on its mainframes. Honeywell, the U.S. distributor of NEC mainframes, for many years sold the Multics operating system for its products, but it decided in early 1986 to abandon Multics in favor of ACOS. Many customers were irritated by Honeywell's decision, since Multics had long been regarded as the operating system most secure from intrusion. ACOS (or GCOS, as Honeywell calls it) does not deliver the same security, but its lower price offsets concerns about security.

Japan has recently broadened its research in operating systems to include the Unix environment. This interest seems to have been sparked by the Japanese recognition that Unix is currently the operating system of choice in most first-rate university computer science centers. Having noticed the growing prominence of Unix in the United States, Japanese researchers have begun working with advanced versions of the system.

Japan is conducting some interesting product engineering in data base technology, but to date this research appears to be largely derivative of U.S. work. The major effort involves the large manufacturers copying IBM-compatible products in the areas of relational data base systems and distributed data bases, but this effort appears to be well behind the innovations found in the United States.

The software factories mentioned above, however, represent a substantial competitive edge for Japan in the area of software engineering. One such firm is the Software WorkBench at Toshiba. The factory employs 2,000 technical people and delivers four million lines of assembly language equivalent code per month. The factory specializes in process-control software for systems such as nuclear power plants, steel mills, and flight guidance. Specialization allows them to achieve a reuse rate of roughly 65 percent; that is, of 3000 lines of delivered code, 1000 lines are new and 2000 lines are recycled from earlier work. The software is quite sophisticated; a typical process-control system might be built around two large mainframe mprocessors, six minicomputers, and a dozen microcomputers. The codes are of extremely high quality; their error rate is 0.3 bugs per thousand lines of code, whereas typical U.S. rates are ten times greater. Much of the software comes with a ten-year warranty; Toshiba will debug any software product during that period at no charge to the customer.

The success of the Toshiba factory is due to careful application of the methods developed by the software engineering research community in the 1970s. Critical methods include the SADT specifications technique, HIPO coding methods, and careful performance estimates. The Toshiba factory uses simple but effective software tools, such as a source code control system, and extensive simulators. Many other Japanese firms have adopted the Software WorkBench concept. Outputs vary, but high-quality, low-error-rate code is common to all of them.

The Japanese practice of tailoring applications to each customer's specification is a fundamental difference between the Japanese and the American approach. The United States abandoned tailored code some time ago. This practice consumes large quantities of programmer time, consequently limiting the resources available for more innovative work. It is also true that software maintenance can consume up to 50 percent or more per year of the original investment. Under these conditions, the Japanese may gain more in software maintenance than they spend in the lengthy process of tailoring their codes.

In contrast to the strength Japan has shown in processor architecture research and development, software research appears to be largely inconsequential. Still, a few areas stand out as enclaves of excellence. As in processor architecture, these areas of excellence tend to be in the product engineering phase.

Artificial Intelligence

Artificial intelligence (AI) is a combination of hardware, systems software, and applications software being developed as a way to allow computers to use heuristics and decision rules in conjunction with sophisticated data bases. The term encompasses a theme common to several different disciplines. These discrete areas of research include speech recognition, natural language processing, and expert systems.

Speech recognition systems translate human speech into useful electronic formats for use in information systems. A related technical problem unique to Asian markets is computer recognition of pictographic characters which are more difficult to encode in a computer's logic than the Roman alphabet. Natural language processing is the keyboard equivalent of speech recognition, or the computer system's ability to interpret direct English- or Japanese-language input without requiring users to convert their needs into the format of an established computer language. An important application of this technology is in allowing machines to translate documents from English to Japanese or vice versa.

Also known as knowledge-based systems, expert systems attempt to create a data base and a set of logical decision rules that together should enable a computer system to perform such functions as medical diagnosis, geological analysis, and computer fault diagnosis. This area of research comes closest to the conception that artificial intelligence allows a computer to simulate the human processes knowledge development and thought. Japanese research in artificial intelligence is also discussed in the Appendix.

Basic Research.   Japan's interest in artificial intelligence stems in large part from the Japanese language. Unlike processor architectures or systems software, artificial intelligence research must contend with the interface between the user and the computer system. This interface often is in the form of computer languages, but it must still employ the user's natural language character set. Americans use Roman characters on their keyboards and screens to communicate with the computer. This seemingly simple interface poses significant problems in Japan, whose written language involves far more characters than any Indo-European languages. This has important consequences in the computer science field. The 8-bit byte so common in all major computer systems can easily handle character sets similar to the Roman alphabet, since 256 combinations of ones and zeros are possible. Even including full upper and lower cases, arabic numbers, and assorted characters such as ampersands, colons, brackets, and the like the 8-bit byte allows computer system designers to develop machines with standardized architectures without regard to the natural language known by the system's users.

The vast number of Japanese characters, however, makes such a system unworkable in that language. It is possible to use architectures employing eight bits to the byte by encoding each Japanese character in two bytes rather than one (using outsize, multishift, or multistroke keyboards). This solution, however, is awkward at best, and the Japanese have long been pursuing a vocal input and alternate character recognition technologies. One of the most important basic research efforts, a ten-year MITI project that started in 1971, cost about 22 billion yen but produced little in the way of concrete results. Today, the country is a major presence in basic research in the field, although the United States is still the leader and is expected to remain so.

Japan is far less sophisticated in basic research into natural language processing. Since U.S. efforts are generally in English, the Japanese would gain little by relying on Western basic research. Nonetheless, research into this field has been sporadic at best, consisting to a large degree of a joint IBM Japan/California Institute of Technology project and ICOT's natural language parsing system.

In the specific area of machine translation of documents from Japanese into English, Japan has achieved some notable basic research success. This is partly due to the country's status as a major trading nation, since Japanese merchants need to have efficient means of getting information to and from their markets. Consequently, at least eighteen different machine translation efforts are active in Japan, spanning virtually all of the major research centers. By contrast, machine translation was largely abandoned in the United States nearly two decades ago, and only a few efforts are currently under way.

The most striking Japanese research in machine translation is currently being conducted through the Mu-project, which the STA created in 1982. Eleven different companies shared 1.64 billion yen in 1984, comparable to the funding level of the more widely known fifth generation project. The Mu-project is using a single translation algorithm which analyzes individual Japanese sentences according to a Japanese dictionary and knowledge of the language's grammar. The resulting Japanese intermediate structure is then converted into an English intermediate structure. This English structure is then reconstituted into a sentence using an English dictionary and a set of transformations intended to remove any remaining artifacts of the Japanese structure.

Several significant obstacles remain before this model can move from the realm of basic research into advanced development. The massive ambiguity inherent in both English and Japanese words implies that much of a word's meaning is actually determined by its context and not by the word itself. The current system cannot accommodate this reliance on context because of the wide variances between the languages.

Advanced Development and Product Engineering. Because basic artificial intelligence research cannot be copied across cultures easily, the Japanese cannot effectively build advanced development efforts on American basic research, as they have in processor architectures. Therefore, the relative status of Japanese advanced development closely resembles that of the country's basic research. In areas in which the Japanese have strong basic research efforts, such as speech recognition and machine translation, advanced development is usually likewise ahead of the United States. In areas where the Japanese have weak basic research, such as expert systems, their advanced development and product engineering is also weak; moreover, because of the lack of basic research, in these areas Japan's advanced development is falling even further behind.

In speech recognition, for example, the strong basic research supports a wide variety of development and product engineering projects, including work by all of the major computer companies. Each company's effort reflects a different focus, as Toshiba, Hitachi and NEC have sacrificed the size of the system's vocabulary in exchange for the ability to recognize speech without regard to who is speaking. Fujitsu, on the other hand, has developed a system with a larger vocabulary and the capability of voice word processing, but each system must be "trained" to understand a single speaker; other speakers must retrain the system before they can use it. NEC has not focused on the large systems its counterparts are pursuing, preferring instead to concentrate on smaller microcomputer-based systems. In the past two years it has marketed a speech recognition system in the United States for use with its microcomputer products. By comparison, American efforts have been fairly ineffectual. About two dozen companies market speech recognition systems, although none have sufficiently large vocabularies to generate much customer interest.

DARPA's Strategic Computing Program has funded some speech research, and IBM has had a voice recognition system prototype operating in its laboratory for several years. The system, which has a comparatively large 5,000-word vocabulary, has attracted much attention but appears unlikely ever to reach production.

Similarly, the Mu-project has led to significant advanced development in machine translation in Japan, again on a par with U.S. development. The United States seems to have a temporary lead in product engineering, however, due to the early efforts of a handful of small companies and one university. LOGOS produces a German-to-English translation system that can be integrated into Wang word processors, and Weidner Communications offers a variety of language pair translation systems integrated with word processors. Weidner is controlled by Japanese interests, however, and is likely to focus on Japanese-to-English technology. Finally, the University of Texas Linguistic Research Center, with funding from Siemens of West Germany, has produced a system for German-to-English translation.

The Japanese are far ahead in the specific area of recognition of Japanese characters. This field appears to have little relevance to American computer system manufacturers, yet Japan is the world's second largest commercial market for computer systems. This market has historically accepted computer systems based on Roman characters because it had no choice; Kanji character recognition was not commercially feasible. As the Japanese develop this technology, however, customers in Japan are likely to begin demanding Kanji capability. If so, American firms could face a significant commercial handicap.

Japanese success in machine translation and language input technology stands in stark contrast to its development work in expert systems. Japan's advanced development and product engineering in this area is far behind that of the United States and falling. In contrast to the Japanese weakness, the United States has seen an explosion of advanced development and new products in expert systems. Much of this research has been funded by DARPA, the National Institutes of Health (NIH), and the National Science Foundation (NSF), and conducted by Stanford University, Massachusetts Institute of Technology (MIT), Carnegie-Mellon University, and SRI International. In addition, private industry has marketed many expert systems products. Teknowledge, for example, has developed prototypes for applications such as consulting and customer education. Similarly, Intellicorp sells software for genetic engineering and the KEE knowledge engineering framework. Digital Equipment Corporation is developing XSel, a system to advise salesmen, and XCon, a system to configure the firm's VAX computer systems.

Communications

A specialized set of hardware and software is needed to enable discrete computer systems to interact with each other, sharing data and programs and allowing users to access data or operate programs on any of several computers from a single terminal. This area will be treated in detail in Chapter 7.

SUMMARY

An overview assessment of Japanese computer science research is depicted in Figure 3–1. Differences in capability between the two countries exist for a number of reasons. The U.S. lead in basic research in all categories appears to be the result of sustained funding of computer science research at major universities over a twenty-year period. Japanese university departments are considerably smaller and contribute to the field primarily through faculty participation in national councils and industrial consulting projects. They are not responsible for advanced training to nearly the same degree as American universities; training as well as research are left to Japanese computer corporations, who do not invest in basic computer science research to the level of the U.S. government.

Japanese weakness in basic research on software is a theme that will be revisited in other chapters. Strong performance in software engineering is the result of a commitment to custom-tailored code and a conscious trade-off decision between investing the extra time required to perfect the code and allocating those resources to more innovative projects. Excellent work in machine translation is a product of economic necessity; the need to understand foreign languages has never had the urgency in the United States that it has in Japan.

Areas where the Japanese are equal to or ahead of the United States are those on which the major computer corporations have focused their internal research budgets and participated in national

Figure 3-1. Overview of Japanese Computer Science Research.

CATEGORY	BASIC RESEARCH	ADVANCED DEVELOPMENT	PRODUCT ENGINEERING
SOFTWARE	− ↗	− ↗	+ ↖
ARTIFICIAL INTELLIGENCE	− ↗	− ↑	− ↑
PROCESSOR ARCHITECTURE AND COMPUTER ORGANIZATION	− ↗	0↗	+ ↖
COMMUNICATIONS HARDWARE	− ↗	0→	+ ↖
SOFTWARE	− ↗	− ↑	− ↗

CODING SYSTEM — JAPAN COMPARED TO USA:

PRESENT STATUS

+ AHEAD
0 EVEN
− BEHIND

RATE OF CHANGE

↗ GAINING GROUND
→ HOLDING CONSTANT
↘ LOSING GROUND

MITI-funded programs. The Japanese industry has committed itself to the advanced development of prototype architecture; successive refinement of those architectures through laboratory experimentation; and high-quality, low-cost product engineering and manufacturing.

4 OPTO- AND MICROELECTRONICS

This chapter examines the scope and quality of Japanese research in non-silicon approaches to microelectronics. This topic was chosen primarily because silicon is considered a fairly mature technology in which future advances are more likely to result from engineering improvements than scientific breakthroughs. The alternatives to silicon, however, are just beginning to be explored, and research in this field is likely to have enormous commercial implications.

The term "microelectronics" refers to digital applications of semiconductors in which the flow of electrons through the semiconductor carries information. This field has been evolving for several decades, beginning with Bell Lab's invention of the transistor in 1947, and continuing through the development of the integrated circuit in the early 1960s. Microelectronics is the building block technology for computers, consumer electronics, and a wide range of commercial and military equipment and instrumentation. Over time, technology and market forces are driving the semiconductor industry towards new chips that are faster, smaller, and cheaper than the current generation. Achieving these improvements depends on exploiting the latest advances in semiconductor technology, which today is based primarily on silicon.

U.S. semiconductor manufacturers have been able to achieve incremental increases in processing speed and addressing capability by improving on the designs of their silicon-based technologies. Each of their new-generation chips is generally compatible with software written for previous-generation chips in the same family. In the near term, these incremental improvements continue to pay dividends at relatively little cost.

In the long term, however, the need to retain software compatibility with a previous generation, and the inherent physical limitations of silicon as a semiconductor material, will prevent the semiconductor industry from making any significant speed breakthroughs without major product discontinuity. Strict adherence to the architecture developed in the early 1980s may prevent the use of more efficiently designed pathways.

The physical properties of silicon, like the architectures used, constrain major performance improvements. The speed of light is finite and partially dependent on the medium through which the light is traveling. The same is true for the speed of electronic pulses. Thus, the performance of any semiconductor is governed by its inherent impedance of electricity and by its density, as measured by the distance a signal must proceed to complete a circuit.

Silicon is more limited than other semiconductors in both impedance and the distance a signal must travel. First, the speed of electricity through silicon is not as fast as it is through other semiconducting materials, such as gallium arsenide. Second, because of this higher impedance, silicon chips emit more heat than other semiconductors. Hence, they must be located farther apart to allow heat to dissipate, and this increases the distance a signal must travel to complete a circuit.

The semiconductor industry is therefore approaching an important threshold. As silicon technology pushes the limits of its own physical properties, incremental increases in speed or addressing capabilities will become increasingly expensive. Moreover, each increment of additional performance requires that the raw material be commensurately more pure. This can lead to an entirely new set of technological problems, as semiconductor synthesizers strive to develop substrates in sufficient quantity and quality. These factors imply that at some point in the near future other materials will need to be substituted for silicon. Supercomputer research is already pushing the limits of silicon as a semiconductor material, and research on gallium arsenide is accelerating.

A wholesale switch from silicon to gallium arsenide is unlikely in the near term. Sophisticated engineering and innovative approaches to silicon technology as well as the high cost of gallium arsenide preclude the demise of silicon in the near future.

The term "optoelectronics" includes applications in which information is carried by photons rather than by electrons. To date, all optoelectronics systems must interface with microelectronic systems in order to be useful; for example, a fiber optic communications system requires microelectronic repeaters which means that at frequent intervals along a channel an electronic component must intercept the optical data stream, convert the information carried by the photons into like information carried by electrons, separate and amplify the signal, and the convert it back into photons for transmission to the next repeater. No pure optoelectronic system now exists to receive, amplify, and transmit optical messages without first converting them into electron-based messages. Finally, in integrated optics, chips may ultimately be designed to handle photons and electrons interchangeably.

JAPANESE ORGANIZATIONS INVOLVED IN III-V RESEARCH

MITI Funding and Private Sector Electronics Research

MITI has played a significant role in the growth of the opto- and microelectronics field in Japan through the funding and organization of several key national projects. One important center for opto- and microelectronics research in Japan is the Optoelectronics Joint Research Laboratory (OJRL) in Kawasaki. This joint laboratory has been funded by AIST and administered by the Optoelectronic Industry Development Association. The structure and operation of the OJRL laboratory are consistent with the description of the joint laboratory system in Chapter 2, with teams of researchers from different companies collaborating on generic technology research. Fourteen companies have participated in research at the OJRL over a period of six years. Representatives from Japanese electronics corporations consulted with each other and with MITI officials on the scope and research agenda for the laboratory before the program began. Thus there has been a well-established consensus as to what is generic technology and what is proprietary. The lab's charter is to investigate opportunities in materials growth and processing, which can be applied in technologies as diverse as optical communications and fast computing. The organization's research interests span a range from the growth of "ideal" large single crystals to such advanced topics as maskless ion beam doping and compound semiconductor device fabrication.

AIST provided roughly $75 million in optoelectronic research funding through 1987. This, however, is only a fraction of national investment in the field. OJRL companies supplemented the MITI funding, and Japanese companies have been conducting proprietary research and development internally. In general, the OJRL has concentrated on materials and closely related problems; parallel research in industry and government labs has focused on complete systems and their overall design.

A separate AIST program (under the Basic Technologies for Next Generation Industries Project) was established in 1984 specifically to examine multiquantum well and superlattice applications. This program, discussed in more detail in the section on superlattices, is administered by the Research and Development Association for Future Electron Devices, involves eleven companies, and will last five years. Each of the AIST programs has specific targets and milestones and is subject to a in-depth progress review after three years.

Research at Nippon Telegraph and Telephone (NTT)

NTT, the national communications company, has supported the largest and most sophisticated electronics research programs in Japan in its Electrical Communications Laboratory (ECL), a large program that consists of 4000 employees at four locations: Musashino, Yokosuka, Ibaraki, and Atsugi. ECL's primary U.S, counterparts are American Telephone & Telegraph's (AT&T) Bell Laboratories and Bell Communication Research. But ECL is three times the size of Bell. It is engaged in a similar scope of work, including optical communications, digital switching, large-scale integrated circuits, and integrated information systems. Because of its large budget and impressive facilities, NTT attracts graduates from the top universities, and the quality of research personnel is extremely high. NTT has taken out over 8000 patents; this is one indication of its contribution to electronic and telecommunications technology.

ECL has an important advantage over its U.S. counterparts that is an important factor in transferring technology from government labs to industry: the laboratory shares personnel with many large Japanese companies. For many years employees of the Japanese computer manufacturers and other large firms have joined NTT for long tours of duty and then returned to their home companies.

NTT's laboratories represent the state of the art in design and construction. The recently completed Atsugi lab consists of several buildings comprising close to a million square feet and employs 400 professional researchers. Research there emphasizes development of silicon-based VLSI, gallium arsenide and Josephson junction technology, and optoelectronic semiconductor devices. It is interesting to note that the Atsugi operation encourages free communication between the silicon and gallium arsenide research groups and transfers personnel frequently between the two. In the United States, by contrast, research teams often concentrate on one semiconductor technology to the exclusion of others. The Japanese have found that advances in silicon contribute to research on gallium arsenide and vice versa. The U.S. approach, moreover, runs the risk that silicon experts will find their skills to be obsolete or that gallium arsenide specialists will have difficulty finding jobs in industry until the material is more commonly used.

Since NTT has no internal manufacturing capability, it makes massive procurements from key Japanese electronics firms. One function of internal NTT research, therefore, is to develop highly detailed specifications for products. The chapter on telecommunications will demonstrate both how these specifications have been an effective barrier to foreign entry into the Japanese market, and how they have also kept Japanese manufacturers from penetrating a number of foreign markets.

Above and beyond its internal research, NTT regularly collaborates with, and contracts research out to, its "family" of supplier firms, many of whom are willing to share proprietary research results with NTT. The special relationship between NTT and its supplier firms has been an extremely effective mechanism for sharing research between government and industry.

Finally, NTT has established national projects in which a number of competing supplier firms participate. In contrast to AIST programs, which result from industry suggestions, these projects are initiated by NTT, which then chooses the participants. In general, NTT either subsidizes research on different approaches to the same problem by different companies or breaks the problem up into pieces, with each competitor working on a different piece. Firms that are successful receive a share of procurement contracts commensurate with their contribution to the research and development process. NTT ensures that all supplier firms ultimately gain access to the manufacturing technology developed during the course of the project.

The NTT Company Law and accompanying Telecommunications Business Law of 1985 stipulated that NTT would become a semiprivate corporation subject to competition from independent service providers. Like the Bell Operating Companies in the United States, NTT will face immediate competition for customer premise equipment sales, value-added network services, long distance, and basic services. NTT shares were handed over to the Ministry of Finance in late 1985; up to two-thirds of them will be sold to the public in phases. Unlike the breakup of the Bell System in the United States, it does not appear now that the company's new status will in any way diminish its role in electronics or telecommunications research and development.[1] NTT's research activities will be discussed again in the chapter on telecommunications.

Japanese University Research on Opto- and Microelectronics

Universities are important contributors to Japanese research in opto- and microelectronics. The Ministry of Education began to support semiconductor interface studies at Japanese universities aggressively in the mid-1970s; the first phase of this research ended in 1978 with the International Conference on Solid Thin Films and Surfaces held in Tokyo in July of that year. This phase can be seen as part of a determined effort by the Japanese to bring themselves up to the state of the art in surface and interface experimental techniques, particularly in high vacuum technology. At about the same time the Ministry committed $60 million to the construction of a synchrotron facility called the Photon Factory, which is used by the OJRL, the universities, NTT, and private industry. Similar facilities (on a smaller scale) have proved valuable in surface and interface studies in the West. Despite this rapid progress, Japanese basic research in many areas of microelectronics lags that of the United States. The Japanese are a long way from building the kind of strong fundamental programs that exist at U.S. and European universities.

EVALUATION OF JAPANESE OPTO- AND MICROELECTRONICS RESEARCH

The remainder of this chapter will assess Japanese capabilities in synthesis of semiconductors, superstructures, optoelectronics, metal contacts to III-V semiconductors, and Josephson junctions.

Synthesis of Compound Semiconductor Materials

Compound semiconductors of the sort required in optoelectronic and microelectronic applications are not typically found in nature; they must be synthesized, or grown, from their component elements. The most commonly used elements are silicon (Si), gallium (Ga), indium (In), aluminum (Al), phosphorus (P), and arsenic (As), in various combinations. These elements are generally combined in several layers—a substrate and one or more applied epitaxial layers. Two principal methods for the synthesis and controlled growth of semiconductor materials are molecular beam epitaxy (MBE) and organometallic vapor phase epitaxy (OMVPE). The latter is also often called metallic-organic chemical vapor deposition (MOCVD). A potentially important technique for producing the substrates on which these epitaxial layers are placed is the liquid encapsulated Czochralski (LEC) method. The MBE and OMVPE methods are used to formulate the epitaxial layers, while liquid encapsulated Czochralski appears to be a promising technique for producing high-quality bulk substrate material.

In all of these areas, Japan has benefited extensively from the availability of results published by leading researchers in the United States. Many second-tier, smaller U.S. firms have also benefited from this research. However, the Japanese government has made a more consistent and substantial commitment to the commercialization of these technologies and has developed detailed plans to achieve this goal. Moreover, U.S. government funding is available primarily for research with specifically military applications. In contrast, Japanese firms regularly apply for research funding (either subsidies or success-conditional loans) to support the development of commercial products. Thus, Fujitsu receives funds for semiconductor laser R&D, which may ultimately be used for high-speed optical data links for its computer systems; Sony receives funds for visible laser R&D, which can be applied in its optical video and audio disk players; and NTT receives funding for short-wave length and long-wavelength laser R&D, which could be used in future optical communication systems.

Molecular Beam Epitaxy. The MBE method for growing compound semiconductors was pioneered by Bell Labs, but several nations have important projects underway. These include the United States, Japan, France, Great Britain, and other European countries. The technology is being developed for application in optoelectronics and in high-speed integrated circuits.

MBE can be used in optoelectronics for both short-wavelength and long-wavelength emitters and detectors, but the research and development required for each are different. The United States initiated short-wavelength (650–870 nanometers) work in the GaAs/AlGaAs materials, and at one point held a significant lead. Most of the U.S. short-wavelength MBE activity has subsided, however, while Japan has pursued the technology with vigor. Fujitsu, NTT, Mitsubishi, the OJRL, and MITI's Electrotechnical Laboratory (ETL) have all conducted significant work using GaAs/AlGaAs materials. In addition, NEC and NTT have also shown impressive results in using AlGaInP/GaInP materials for use in visible light emitters and double heterojunction lasers.

In long-wavelength optoelectronic MBE research, Bell Labs was again an important early participant; it is still conducting projects in this area. Other major organizations involved in long-wavelength MBE research include British Telecom in the U.K. and NTT, the Tokyo Institute of Technology, and Sumitomo Electric in Japan. AT&T Bell Laboratories still holds the lead in this field, primarily due to its successful MBE growth of InGaAsP. Other researchers working toward growth of this material have been stymied by their inability to control the simultaneous incorporation of As and P.

MBE is also used for the development of high-speed device and integrated circuit applications. Japanese and American research teams are on roughly equal footing here, using AlGaAs/GaAs and InAlAs/

InGaAs materials. Major participants in the U.S. include AT&T Bell Labs, Rockwell, TRW, Hewlett-Packard, and IBM. In Japan, the leaders are Fujitsu, NEC, and Hitachi. The U.S. labs can generate better speed and frequency results, while the Japanese have developed more complex integration schemes.

Organometallic Vapor Phase Epitaxy. Whereas MBE is used extensively for both optoelectronic and high-speed device/integrated circuit applications, OMVPE is currently used primarily in optoelectronics for the growth of such III-V compounds (so named because they come from the third and fifth column of the periodic table of elements) as GaAs, AlGaAs, InP, InGaAs, and InGaAsP. The United States generally is leading these research efforts, through teams at Xerox's Palo Alto Research Center, Hewlett-Packard, and Cornell University. The gap between U.S. and Japanese work is narrowing, however, as Sony, Hitachi, and the OJRL, among others, have reported impressive results. Of particular note is Oki Electric's pioneering growth of GaAs on silicon substrates. This work seems to be outpacing similar work at MIT's Lincoln Laboratory.

Liquid Encapsulated Czochralski GaAs. LEC produces most GaAs substrates now being used throughout the world. Extensive research is currently under way in both the United States and Japan; the OJRL and Sumitomo Electric are leading the Japanese effort, while Westinghouse, Rockwell, Texas Instruments and Hewlett-Packard are at the vanguard of U.S. research. Other nations are also developing LEC technology; Cambridge Instruments (United Kingdom) and Cominco (Canada) are the most notable participants. The United States and Japan are in comparable positions, but Japan seems in a better position to commercialize this technology. In particular, Sumitomo Electric's semi-insulating In-doped GaAs substrates appear on the verge of commercial production. Japan remains the only country with a long range objective and an executable plan for dominating this technology.

Superstructures

Combinations of several layers of semiconductors can be created to form superlattices or multiple quantum well structures. These superstructures will ultimately use high electron mobility transistors (HEMTs) and three-dimensional structures for potential use in artificial intelligence systems where high speed and density of components is critical.

Superstructures using AlAs/GaAs and AlGaAs/GaAs have been investigated in great detail. The close match of the component elements in these semiconductors has enabled researchers to construct layers as thin as 100 nanometers; because this thickness is smaller than GaAs's electron wavelength, size is now controllable at the quantum level. The new research and product development possibilities engendered by this quantum control of the semiconductor layers, and the corresponding possibility of transmitting signals electronically as well as by light, have elicited an enthusiastic response in Japan. In the United States, by comparison, no similar commitment to superstructure research seems evident.

Japan has enunciated several major objectives in its superstructure research. First, it is spending about 1 billion yen annually to produce ultrafast integrated circuits using HEMTs. Second, it is funding research totalling 1.8 billion yen on three-dimensional integrated circuits that can be used in artificial intelligence applications. Other programs, such as one to develop interchip and intrachip connections, are also under way. In total, MITI (through the AIST Basic Technologies for Next Generation Industries) has funded about 25 billion yen for superstructure research.

In the area of ultrafast integrated circuits, Fujitsu has shown impressive results in its HEMT research. Its level of sophistication and integration has increased about twice as fast in HEMT technology as it has in silicon technology, primarily because Fujitsu sees important potential commercial applications. Specifically, HEMT is viewed as a prime candidate for memory units in Fujitsu's future supercomputers and as a computational element in some smaller computer systems. HEMT has favorable properties with regard to miniaturization and heat dissipation, making it potentially useful in office-based devices such as computer graphics systems that require intensive computation in real-time environments.

Following the logic of a paper initially published in 1965 by Bell Labs vice president Jack Morton, the Japanese hope to use this HEMT research to create fully functional devices—not merely single-purpose components—at the semiconductor level on three-dimensional chips. Such fully functional devices are beyond the reach of conventional silicon-based computer systems. The vast complexity and computational requirements of a full-fledged artificial intelligence system would require the integrated operation of many thousands of integrated circuits operating in conjunction with one another. Yet because of silicon's heat dissipation problems, these circuits would have to be spread over a relatively large area; hence the time required for pulses to complete their connections would rise astronomically, prohibiting any real-time use of such a system. By comparison, an artificial intelligence system embedded in a single superstructure, such as an HEMT three-dimensional chip, would have the double effect of "hard-wiring" many functions into the semiconductor, thereby reducing the size and complexity of the application, and of enabling far more transistors to operate in close proximity without emitting too much heat, thereby immensely improving speed.

Primarily due to MITI's commitment and Fujitsu's capability, Japan has a firm lead over the United States, in superstructure research, and the gap will probably widen. Still, some superstructure research has begun in the United States. AT&T Bell Laboratories, IBM, and some Defense Department installations are conducting the most prominent efforts, but they lack the kind of targeted government support so evident in Japan. Because neither the government nor the full breadth of the U.S. semiconductor research community recognizes the potential commercial importance of superstructure research and the possible development of fully functional devices within a single semiconductor, the United States is likely to lag Japan in this area, even though the pioneering work was first done in American laboratories.

Optoelectronics

In contrast to the theoretical orientation of semiconductor synthesis and advanced semiconductor structures research, optoelectronics research and development are product-oriented. Current applications in the field of otoelectronics include semiconductor lasers, light-emitting diodes (LEDs), light wave communications, and optical storage.

Optoelectronics are a big business in Japan. In 1982, optical communications (through fiber optic telephone cable) alone represented an $85 million domestic Japanese market; local area networks using optical fibers for computer systems accounted for another $70 million. Lasers generated $5 million in sales, and LEDs used in communications totaled $10 million. The market has been booming, with sales increasing tenfold over the past five years. Worldwide optoelectronics sales are expected to reach $100 billion in annual sales by the end of the century.

The Japanese optoelectronic market can be divided into three basic areas: components, equipment, and systems. Components, which represent half the total market, include optical fibers; lasers and LEDs; and photodetectors. Japan is well ahead of the United States in this area and is pulling away. Equipment using these components includes optical disk drive units, laser printers, and electronic fields. This equipment can be integrated into telecommunications and local area network systems. The United States still holds a lead in equipment and systems, but Japan is closing the gap.

Japanese firms have begun extensive research and development programs in lasers, integrated circuits and optics, light wave communications, and optical storage. Funding for this research comes from several MITI programs (including the OJRL), NTT's Information Network System and from corporate sponsorship of proprietary programs.

Solid-State Lasers. The specialty of semiconductor lasers is among the fastest-growing within the optoelectronics industry, at 28 percent annually. It has received a growing amount of support in both the United States and Japan, along with the United Kingdom, Germany, and France. The U.S. effort dates to about 1969, while the Japanese did not begin serious research until 1976. Consequently, the United States has held a lead in this field, but the gap is narrowing rapidly. Both nations are far outpacing efforts elsewhere.

Japanese progress is in large measure the reflection of a joint decision by MITI and Japanese industry to target this area for intensive research and development. AIST reported in 1981 that it hopes to capture 40 to 50 percent of the optoelectronics market over the long term. To achieve that goal, MITI has provided over $100 million in funding for research in solid-state lasers since 1978. (This figure, impressive though it may be, is considerably smaller than the $430 million MITI is spending on the fifth generation project over a ten-year period.) Because this funding has been consistent over an eight-year period, Japanese research adhered to a coherent technical agenda path and has not been subjected to shifts in technical direction or level of effort due to changes in the marketplace.

The Japanese research in solid-state lasers falls into several categories. Prominent among these approaches is the nation's commitment to gallium arsenide as the substrate of choice for laser technology. Japan has decided that it will be the worldwide leader of GaAs substrates. In 1982, Japan began construction of a massive manufacturing capability for these substrates. In the period from 1982 through 1985, Japan increased its bulk GaAs production capacity fourfold to eight tons, which could be used for 10 billion lasers. Sumitomo alone expects to make a hundred tons per year by 1988 using gallium from the People's Republic of China. The effort has paid off—the only reputable commercial sources of GaAs substrates are in Japan, so American firms must either grow their own material or rely upon Japanese suppliers.

Whereas the Japanese program has been spread among the OJRL, NTT, Sumitomo, and many other firms, the U.S. effort has consisted almost entirely of work done at AT&T Bell Laboratories. This work is not directly comparable to the Japanese research. Bell appears more interested in pursuing simpler and more reliable methods of fabrication, even if these simpler processes may adversely affect device performance. The Japanese, by contrast, have tended to work on sophisticated structures to obtain better device performance, suffering through what they see as temporary manufacturing difficulties. These work patterns are correlated with different national strengths. Specifically, the United States has been responsible for most of the conceptual breakthroughs in the field, thereby providing revolutionary improvements in fabricating techniques. Japan, by contrast, has led in structural inventiveness, which has enabled it to generate major incremental improvements on existing technology.

Light-Emitting Diodes. Since MITI turned its attention to semiconductor laser research, Japanese interest in LEDs has declined dramatically. One evidence of this shift is that the number of papers published in LED research changed very little between 1976 and 1982, whereas the number of papers published in semiconductor laser research grew by about 200 percent. LED research and development can be considered mature, with the United States and Japan operating at parity. As newer and more advanced technologies are brought into the research labs, the funding for LED research is likely to taper off significantly. The field of integrated optics has been uneven. A worldwide decline in interest in the field which began in 1981 is still not well understood. Japan, in any event, has placed considerably less emphasis on integrated optics, favoring instead research on semiconductor lasers. The United States appears to place approximately equal emphasis on both.

In this instance Japan is taking a risk by putting all of its eggs in one basket. MITI's shifting of research funding and the resulting changes in research activity suggest that government and industry see solid-state lasers as the primary means toward the ultimate development of pure optoelectronic systems. If this guess proves correct, Japan will be in a strong position to dominate a booming market for solid state lasers. If, however, another technology eventually becomes critical to the development of wholly optoelectronic systems, Japan will find itself far behind the United States and other nations.

If the Japanese approach entails some risk, so does the U.S. approach. The U.S. effort is so diffused among different technologies research that it may be difficult to build the critical mass needed to achieve superiority in any one area. If semiconductor lasers are ultimately bypassed in favor of some other technology, the United States may not be able to capitalize on this development because too few American researchers will have explored the prevailing technology. If semiconductor lasers are indeed the wave of the future, American research will be hard-pressed to equal the productivity of the targeted Japanese research.

Metal Contacts to III-V Semiconductors

One area in which the United States has maintained parity with Japan is in the art and technology of metallizing GaAs circuits. The successful commercialization of GaAs high speed devices depends upon a useful and high-yield technology for producing metal contacts that interconnect discrete GaAs circuits. Japan probably has a slight lead in high temperature stable Schottky barrier gate metallurgy. The W-Si and Ti-W-Si gate metallurgy developed in Japan has found wide application. The Au-Ge-Ni ohmic contact was invented in the United States and is also widely used. However, most researchers agree that this technology will not be valuable for large-scale integration (LSI) applications because of a low eutectic temperature.

The United States leads in developing better and more thermally stable ohmic contacts. Furthermore, surface science in the United States and Europe is starting to result in important contact innovation—specifically in new concepts for Schottky barrier contacts. The United States is in a good position to take advantage of its knowledge in this area. This can be accomplished by involving companies who are not active in GaAs technology but who have related expertise and experience. The Japanese do not yet have a complete contact technology suitable for the high yield manufacture of LSI circuits.

Josephson Junctions

In contrast to semiconductors, Josephson junctions are based on superconductive metals. Superconductive interconnections can make use of lossless and dispersionless transmission lines connecting fast switching binary digital devices with a minimum of cross talk. The Josephson junction, in its most common form, consists of two superconducting films separated by an insulating layer less than 100 angstroms thick. Conduction current can be transferred across the insulating barrier through two channels, one binary and one continuous. First, single electrons can tunnel from one electrode to the other in a normal manner that can be modeled as a nonlinear (e.g. binary) function. The second is the Josephson tunneling of superconducting pairs of electrons, which is a linear function in which current oscillates across the junction at a frequency proportional to the voltage applied.

Signals propagate across series of such junctions on superconducting transmission lines located over a superconducting ground plan. These lines are nearly lossless and dispersionless, while the ground plane confines the electrical and magnetic fields so that cross talk between adjacent lines is negligible. While these capabilities provide Josephson junctions with some advantages over semiconductor-based circuits, they also impose some important inconveniences, such as low gain, a lack of a convenient inverter, and latching gates. The most important impediment, however, is that Josephson circuits until recently must have been kept at -269 Celsium (-453F), the temperature of liquid helium. With the advent of high temperature superconductors, the circuits can be kept at liquid nitrogen temperature, -196 Celsium (-314F). This is a major breakthrough and can change the outlook for Josephson junction technology in a major way.

The primary advantage of Josephson technology is its potential application in ultra-high-performance computers. Several attributes are relevant in this respect. One is speed; individual junctions switch from zero voltage to nonzero voltage in a time determined by the product of the junction capacitance and the shunt resistance. This time is typically on the order of ten picoseconds, far faster than semiconductors can switch.

A second attribute is power; whereas the performance of semiconductors is limited by their heat emissions (and, therefore, cooling requirements), Josephson junctions operate at very low levels of power dissipation. Common Josephson devices operate at typical voltage levels of about 2 millivolts and dissipate perhaps 2 microwatts per gate. Tens of thousands of such gates per square centimeter are effectively cooled by immersion in liquid helium or for high temperature superconductors in liquid nitrogen.

A third feature is packaging. The package of a circuit generally contributes two-thirds of the total delay in typical high-performance machines, including time-of-flight delays, charging times, and package loading delays. By comparison, Josephson technology allows for a three-dimensional package denser than semiconductor packages. This improves speed by reducing intercircuit distances. Second, Josephson technology's superconducting transmission lines permit packaging with only one level of wiring in each of two dimensions, simplifying the construction of the integration.

The Josephson effect has attracted significant research interest since its discovery, primarily among theoretical research scientists. The commercial possibilities of Josephson technology have also been recognized for some time, although the successful exploitation of this technology for commercial use is limited by the expense of development. The first and largest commercial development program was initiated in 1967 at IBM, where it represented the first serious attempt to supplant semiconductors with an alternate integrated circuit technology. The IBM effort, which cost about $20 million per year, recognized that the potential advantages of Josephson technology would have to be enormous on a system-side level, since the technology would require the development of an entirely new hardware system architecture.

IBM's activity attracted the attention of rival firms, such as Bell Laboratories and Sperry (now Unisys) in the United States and NTT in Japan. The Bell and Sperry excursions were fairly minor, but the Japanese interest was intense, spurred by MITI's recognition that

Josephson devices could be used in a broad range of applications beyond high-speed computing. Such nondigital applications could include analog signal processing, microwave amplifiers, millimeter and microwave detectors and mixers, infrared detectors, and analog-to-digital converters. The Josephson effect is also used by the U.S. National Bureau of Standards as to define the volt. Still, it is the digital use of Josephson technology that represents the largest commercial market and the most significant international competitive threat.

Despite the many potential commercial applications of Josephson technology, IBM decided in late 1983 to terminate its Josephson junction computer development program and retain only a small basic research effort. The company explained that the potential advantages no longer outweighed the cost of developing the entirely new hardware systems needed to exploit the technology. Three specific problems were cited. First, IBM could not fabricate the minimum 4K cache memory required for future systems without redesigning key circuits. Second, such a redesign would require an additional two years of development; IBM expected that in the intervening two years silicon technology would likely close the performance gap significantly, to the point that Josephson junctions would operate only about twice as fast as silicon. IBM believed that the cost of developing a Josephson computer could only be justified by at least a tenfold performance improvement. Finally, IBM believed that the cache problem foreshadowed a similar problem in fabricating system memories that would be needed in subsequent stages of the system's development.

While the difficulties imposed by Josephson technology caused IBM to cut back its program, the Japanese have retained and redoubled their efforts at exploiting this technology. Several distinct Japanese efforts are currently under way; three of these are industrial programs sponsored by MITI, NTT, and Mitsubishi. There are also several university programs sponsored by the Ministry of Education and Culture. These programs are not coordinated in any explicit manner, but they do seem to be mutually supportive through information exchange resulting from technical journals and meetings of professional associations.

The MITI Programs. Four distinct programs are currently in progress under MITI's auspices, at the Electrotechnical Laboratory, Fujitsu, NEC, and Hitachi. These programs are being funded through the National Super Speed Project (NSSP), which allocates about 27 percent of its resources to Josephson technology. (The remainder is split three ways: gallium arsenide, 27 percent; high electron mobility transistors, 27 percent; and software/architecture, 20 percent.) These programs are not expected to reach commercial fruition for several years; devices may be realized in a couple of years, but complete systems are not likely until the 1990s.

Because of their potential overlap, the MITI programs are atypical in several operational respects. The Japanese government, not contributing companies, will own title to any patents developed during this research and will license the patents according to its own interests. To overcome potential problems in transferring technology among the distinct projects, each researcher at the industrial MITI-supported projects has been sent to AIST's ETL for an extended assignment.

ETL's research has been primarily concerned since its inception with the process of manufacturing Josephson circuits, particularly through the use of all-refractory device technology. (IBM, by comparison, used lead-alloy technology.) ETL is particularly interested in increasing the packing density of Josephson gates and in ensuring their reproducibility in large volumes. The laboratory has innovated the use of niobium nitride rather than pure niobium as the refractory material in order to exploit that compound's larger superconducting energy gap. ETL is also making notable progress with logic and arithmetic functions, having demonstrated an 8-bit adder with a delay of 400 picoseconds and power consumption of 200 microwatts, as well as a 4×4 multiplier with a propagation delay of one nanosecond. These results are ten and five times faster, respectively, than can be achieved using current gallium arsenide technology.

The clear sense of purpose evident in ETL's desire to improve the manufacturability of Josephson junctions is absent at the largest MITI-sponsored Japanese program, the Fujitsu effort. Because the company's strategy has been to adhere strictly to IBM mainframe architecture, Fujitsu has been unhappy about pursuing Josephson technology in the wake of IBM's withdrawal. The company is far smaller than either of its primary competitors (Hitachi and NEC), and it believes that its interest in remaining dominant in HEMT precludes any significant investment in Josephson technology. In keeping with its adherence to IBM technology, Fujitsu's emphasis has been on the reliability of lead-alloy circuits, in contrast to ETL's choice of niobium nitride. Some evidence exists that the firm is switching to all-refractory material, however, both to follow ETL and because it had run into significant problems with lead-alloy material.

Like Fujitsu, NEC is in the process of switching from lead-alloy to all-refractory material, using a niobium full-wafer process. The company has concentrated on the fabrication technology, but its effort is small: NEC has assigned only nine professionals to the project, versus twelve at Hitachi and nineteen at Fujitsu. However, the company seems committed and enthusiastic and has promised to pursue the technology even if MITI withdraws funding.

Of all the MITI programs, Hitachi's is the most resistant to all-refractory technology. The firm is exploring novel circuit designs using lead-alloy material, following work pioneered at IBM. The company has developed a 600-picosecond memory chip, but it is unknown at present whether the chip actually can be fabricated.

The NTT Program.   Following IBM's departure, NTT has the largest Josephson program in the world, with twenty-five professionals working on fabrication and circuit design at the ECL; an additional thirteen professionals are investigating non-digital applications of Josephson technology. NTT's Atsugi lab has refined the fabrication of lead-alloy devices such as 1K memory chips to a high art form. A critical path through the memory has been operated, but it is not clear whether fully functional chips have been tested at Atsugi. The work is a direct extension of IBM's initial efforts, but its focus on memory, compared with IBM's focus on logic, will presumably give it some leverage in circumventing the problems IBM encountered with its cache circuit fabrication. The wide scope of NTT's work is underscored by its impressive staff: In addition to the expertise tackling the memory problem, NTT has some of the best Josephson logic circuit designers to be found anywhere.

NTT has complemented its memory and logic work with some advanced sampling and packing research. While the technology at first blush resembles IBM's abandoned work, its performance is in fact better than anything IBM had ever published, in terms of rise time and measured crosstalk, but only because it spaced the chip's pins farther apart. This allowance of greater pin distance is indicative of NTT's policy of tackling intermediate, accessible goals in sequence rather than setting more ambitious initial goals. The company has also shown significant progress in developing refractory materials, understanding junction tunnel barriers, and exploring the properties of Pb-Ba-Bi-oxide, an unusual superconductor.

Other Programs. Mitsubishi has a small program sponsored by MITI, concentrating on digital applications of Josephson junctions. The six-person group appears capable of pioneering work, but lacks both the equipment and the manpower to engage in anything but lead-alloy technology. Mitsubishi is likely using the group more to scout the surrounding territory than to conduct original research.

Universities have also mounted Josephson technology research, but much of it is theoretical in nature and not likely to lead to any commercial applications in the near future. These programs seem inferior to comparable American programs, however.

With the recent discovery of high-temperature superconductors, considerable reemphasis is being placed on Josephson junction technology, in Japan, the United States, and elswhere. Having continued their efforts, unlike IBM, the Japanese are clearly in a very good position to capitalize on their investment. The competition in this area is fierce, and the winner can potentially dominate the future supercomputer market for years to come.

American Research. Josephson technology in the United States continues despite IBM's withdrawal, but at a diminished level; major programs exist today only at a few universities, Department of Defense installations, and some small, new ventures. The university research programs, unlike those in Japan, play an important role in the development of digital applications for Josephson technology—in large part because these programs enjoy a friendly and unusually cooperative relationship with industrial scientists. By far the most prominent academic program is at the University of California at Berkeley, which has developed both lead-alloy and niobium technology. The group has contributed a wide variety of theoretical and practical circuits and has often attempted to explain the potential significance of alternate approaches rejected by IBM. Other important university programs are at Stanford and Cornell.

As valuable as the research programs are, they cannot completely substitute for industrial development, since several important questions seem appropriate for study only in industrial labs. These include manufacturing methods that might require almost yearly updates in equipment, extensive CAD tools, and analyses of projected margins and yields of various circuit designs.

The Defense Department is pursuing some research along these lines, which can to some extent cross-fertilize commercial development. For more than a decade, the Navy has been investigating refractory materials to improve the reliability of Josephson devices, a goal similar to ETL's. In particular, the Navy is interested in the continuous alternating current (i.e. analog) feature of Josephson technology, since this technology can be used in the Superconducting Quantum Interference Effect Devices (SQUID) program for detecting submarines. The Naval Research Laboratory has been influential in guiding private sector research at companies like Sperry, which switched from lead-alloy to niobium nitride technology under Naval tutelage. While the Navy's interest is in analog applications, Sperry has recognized the potential digital applications Josephson technology has for its information systems business.

The final major avenue of Josephson junction research in the United States has been the entrepreneurial startup companies. One such venture is Hypres Inc., founded by Sadeg Faris, who was instrumental within IBM's Josephson research both as a contributor and as a critic. The company remains in the development stage. A second venture, Micrilor, was founded by researchers formerly associated with MIT's Lincoln Laboratories to pursue superconducting electronics in the signal processing business.

The discovery of high-temperature superconductors has had a major impact on U.S. technology. All key players in the field are rethinking their investment and developing new plans and strategies. The U.S. government is also funding research in this area. Only a few months after the discovery, for example, DARPA launched a major initiative focused on military applications for these superconductors. Simple devices have already been built from the new materials, but the technology is still too young to have manifested all of its possibilities.

SUMMARY

Japanese research in non-silicon-based opto- and microelectronics is summarized in Figure 4-1. In this technology, as in computer science, the United States leads in many areas of basic research, espe-

Figure 4-1. Overview of Japanese III-V Research.

CATEGORY	BASIC RESEARCH	ADVANCED DEVELOPMENT
MOLECULAR BEAM EPITAXY	– ↑	0 ↑
MULTI-QUANTUM WELLS AND SUPERLATTICE STRUCTURES	– ↑	0 ↑
SEMICONDUCTOR LASERS LEDS AND INTEGRATED OPTICS	– ↑	+ ↑
AVALANCHE DETECTOR AND OPTO-ELECTRONIC INTEGRATED CIRCUITS	– ↑	– ↑
METAL CONTACTS TO III-V SEMICONDUCTORS	– ↑	– ↑
JOSEPHSON JUNCTIONS	– ↑	0 ↑

CODING SYSTEM — JAPAN COMPARED TO USA:

PRESENT STATUS

+ AHEAD
0 EVEN
– BEHIND

RATE OF CHANGE

↑ GAINING GROUND
→ HOLDING CONSTANT
↘ LOSING GROUND

cially in an understanding of the scientific properties of gallium arsenide. This lead is the result primarily of an early start in the field. An understanding of the science and device physics of GaAs has not been translated into notable commercial success, largely because very few U.S. companies viewed GaAs as an important semiconductor material in the future. With the notable exception of Cray and a few other firms, American industry seems tied to silicon, preferring to continue developing relatively minor, incremental improvements in existing technology. Figure 4-1 rates the United States and Japan even in Josephson Junction research even though the text suggests a Japanese lead, because the interest in superconductivity has generated renewed U.S. activity in this field since the JTECH report was published.

The United States attitude toward III-V research might be described as "get-rich-quick-and-get-out." U.S. scientists have been responsible for major breakthroughs such as Cr-doped semi-insulating GaAs, MOVPE, MBE, GaAs injection lasers, and hetrojunction bipolar transistors, among others. Yet American research has been characterized by periods of frenetic research and rapid publication by the basic research community, followed by inattention until Japanese technology has superseded the original American efforts. By that time, however, the academic and industrial researchers responsible for the initial U.S. expertise have moved to other projects. The early expertise gained by companies like Texas Instruments, RCA, General Electric, and even IBM has largely been lost, leaving the U.S. field dependent upon a work force trained within the last decade by a few select universities: Stanford, MIT, the University of California at Berkeley, Cornell, and University of Illinois. Government support for opto- and microelectronics in the United States is targeted at programs with clear military applications.

The Japanese have taken a longer-term view and acknowledged the inherent limitations of silicon; namely, that it does not emit light and is inefficient with respect to heat generation. The industry as a whole is committed to developing new materials. This industry consensus is reflected in MITI funding for joint laboratories and other collaborative projects as well as in work at ETL and NTT's ECL. Government funding is complemented by investments in research and development by leading Japanese electronics firms and is used for projects with commercial rather than military applications.

The Japanese have benefited from early U.S. breakthroughs by studying the technical literature carefully and building on it. They have managed to maintain a consistent applied research and engineering effort, particularly in areas such as Josephson junctions, despite the uncertain long-term payoffs.

NOTES

1. Before the 1984 breakup, the Bell System was the official name for the American telecommunication company. It included the American Telephone and Telegraph (AT&T) company, Western Electric Corporation, and Bell Laboratories. After the court-ordered breakup, Bell Laboratories split into two parts: AT&T Bell Laboratories, under AT&T's control, and Bell Communications Research, a research and development arm of the divested regional telephone companies.

5 ADVANCED POLYMERS

The development of many forms of advanced materials began in earnest following World War II. Between 1955 and 1970 the industry grew rapidly in both size and strategic importance, as U.S. firms produced an almost explosive development of new materials. Several factors spurred this growth. With the termination of its Rubber Reserve Program in 1955, the Department of Defense shifted its attention toward improved high performance materials for advanced military applications. When the USSR launched Sputnik in 1957, it created a new challenge which resulted in the establishment of the National Aeronautics and Space Administration (NASA) and the development of the related aerospace industries. Finally, the plastics revolution pioneered by leading chemical companies promised to produce a virtually endless variety of new materials for numerous potential applications. These materials consisted of polymers (including plastics), metals, metal-plastic composites, silicon-carbide, boron, and ceramics.

This boom lasted about a decade. In the late 1960s, the priorities of several key participants began to change. The materials industry, more concerned about current profits, and financial stability, reduced spending on research and development. At the same time, the Department of Defense began to emphasize the need for prototype development of technologies discovered during the basic research it had supported.

By the early 1970s, the materials field was believed to have matured. Industry experts felt that more opportunities lay in understanding and improving existing materials than in developing new ones. Roughly a decade later, several of the more vertically integrated industries in the United States, notably the electronic firms, expressed the need for new and improved materials. Some of these companies have established research programs to develop new proprietary materials tailored to provide specific performance attributes required for their product lines. These new materials require only a short time between conception and production, since implementation is done internally. Moreover, the large corporations sponsoring this internal research can typically spread the development cost over a large production volume, thereby reducing the per-unit cost and inducing the approval of ever more intensive programs. The rapid development and relatively low per-unit cost of recent materials research is especially prominent in certain segments of American industry.

In Japan, the early history of materials development followed a different course. During the period of intensive American research, between 1955 and 1970, the major Japanese chemical companies either licensed technology from the United States or set up joint ventures to manufacture most of the proven, commercially available commodity and specialty materials. These companies then were able to improve these products and the processes used to produce them. In the 1970s, the Japanese industry licensed the improved processes back to the United States.

After the Japanese had harvested most of the gains possible from the commodity and specialty polymer technology they had imported a decade earlier, they recognized the need for new sources of proven technology. Neither Japan nor the United States had yet developed and sufficiently tested certain new polymers, and Japan began to license new, unproven technology from the United States. Japanese chemical companies are still working on these materials, but so far they have not been successful in bringing them to the marketplace. U.S. companies typically abandoned research on these materials once they decided they could not identify a target market for them.

Through the 1970s and 1980s, Japanese and American materials research efforts have operated within similar constraints, but have followed divergent paths. The chemical companies in both nations were severely affected by the Organization of Petroleum Exporting Countries (OPEC) price shocks and by two global recessions. Nevertheless the Japanese continued during that period to establish modern, well-equipped research and development centers. Moreover, through the Ministry of International Trade and Industry (MITI), Japan has articulated a national strategy intended to develop a base for technological leadership in advanced materials through the 1990s. In the United States, by comparison, no coherent national strategy exists for a broad research effort in advanced materials.

Another significant difference lies in the Japanese industry's approach to growth. Japanese chemical companies are increasingly concerned with improving productivity as a way of funding research and compensating for Japan's lack of raw materials. For example, Sumitomo Chemical now generates $500,000 in revenue per employee, four times the level achieved by an American chemical company of similar size. The company reduced its work force (and therefore its labor expenses) by more than 50 percent, presumably through use of automation and robots in chemical processing.

Between 1980 and 1985 Sumitomo also spun off eight affiliates, in the belief that smaller, cost-efficient units would be more competitive than one large, centralized company. The affiliates act as autonomous business ventures whose management can make timely decisions about research, production, and marketing. This trend stands in stark contrast to the U.S. industry, where the emphasis on ever-increasing plant size has pushed companies toward acquisitions, mergers, and centralization.

Rather than attempting to encompass the range of new materials research now ongoing in Japan and the United States, this chapter focuses specifically on advanced polymers. This choice is based in large measure on the projected commercial implications of continued Japanese progress. The use of polymers in electronic applications alone represents a several billion dollar business and is likely to reach $10 billion by 1990. The continued technological development of computers, aerospace vehicles, and other products is certain to demand significant improvement in the performance characteristics of polymers in a variety of applications.

JAPANESE ORGANIZATIONS INVOLVED IN POLYMER RESEARCH

MITI Support and Private Sector Polymer Research

The development of new materials figures prominently among Japan's technological goals over the next decade. The AIST Program on Basic Technology for Next Generation Industry incorporates seven projects on new materials. Of these seven, five are on advanced polymers, one is on ceramics and one is on metal alloys. Government support in the United States for research on advanced materials is almost the reverse, with a far greater commitment to metals and ceramics versus polymers. The seven areas targeted by AIST will run for eight to ten years. Fifteen of the sixteen Japanese national laboratories are involved in advanced materials research, although only the Research Institute for Polymers and Textiles at Tsukuba Science City focuses exclusively on polymers. This lab, with 104 scientists, is somewhat smaller than the others, but it is perceived as carrying out high-quality work on relevant topics. It is especially prominent in synthetic membranes and high performance plastics.

MITI has also recently established three new materials labs for polymers, ceramics, and metals, with the intent of providing materials research support to new companies. These labs will be staffed with 50 to 100 scientists each.

Polymer Research in ERATO

The ERATO program described in Chapter 2 includes one project on specialty polymers known as the Ogata Fine Polymer Program. The effort consisted of twenty-five to thirty people and was directed toward high-temperature monolayers, separations of optical isoners, conducting polymers, molecular particles, and absorption systems. It appeared in 1986 that good progress had been made in all five areas.

Polymer Research in Japanese Corporations

Japanese chemical companies invest heavily in research and development, in both human and financial resources. For example, 25 percent of Sumitomo's employees are in research and development, whereas the percentage for American companies tends to be significantly lower. It is also noteworthy that between 1979 and 1983, Sumitomo's research budget doubled even though sales were flat and the company was suffering through a severe recession in petrochemicals. Specific achievements of individual companies will be discussed later in the chapter.

University Research on Polymers

A consistent pattern in the development of Japanese polymer technology is the lack of indigenous invention or effective conceptual university research. The absence of strong basic research is likely to be a problem in the future. The Japanese are aware that they have depended heavily on U.S. technology in the past, and that since much of the advanced research in this field in the United States has been greatly curtailed, it is no longer available as a resource. Thus it is incumbent on Japan to establish its own infrastructure in basic materials science and to train scientists who can translate this science into useful new materials.

EVALUATION OF JAPANESE ADVANCED POLYMER RESEARCH

The JTECH Panel on Advanced Materials examined high tenacity/ high modulus polymers, polymers for electronic applications, gas and liquid membrane separation, and biopolymers. Their findings on the quality and direction of Japanese research in each of these areas is summarized here.

High Tenacity/High Modulus Polymers

The term "high tenacity/high modulus" (or "high T/M") refers to specific performance attributes of a group of polymers rather than to any structural or inherent common properties. Materials in this class, such as aramids, polyimides, liquid crystalline polyesters, and ultra-high molecular weight polyethylene, can be fabricated into strong, stiff fibers, films, and self-reinforced plastics and moldings for use in a variety of military and commercial applications. These polymers, most of which were discovered and initially developed in the United States, exhibit exceptional strength (tenacity) and stiffness (modulus). In addition, they are also exceptionally stable even at high temperatures; many have melting points above 500 degrees Celsius. Yet their coefficients of thermal expansion are generally low, enabling them to be used in a variety of applications in fibers, plastics, and films. For example, fibers can be used on oil rigs for ropes, in ballistic vests as protective clothing, and in tires and hoses as rubber reinforcement. Similarly, plastics can be used in self-reinforced moldings, precision parts, and thermally stable moldings for aircraft. Finally, film applications include insulation of electrical and electronic apparatus.

Most materials in this classification do share important structural similarities at the molecular level. In particular, these materials generally have an extended chain configuration of the constituent polymeric molecules. The component molecules are of high molecular weight and have a high degree of molecular orientation. The primary research in high T/M materials was conducted initially by the U.S. chemical companies, which produced several important commercial products. DuPont, for example, introduced Kapton polyimide film in 1968 and Kevlar aramid fiber in 1975, while Celanese brought out Vectra polyester resin in 1985. Major groups of high T/M materials include aramid fibers, polyimide, liquid crystalline (LC) polyester, and ultrahigh molecular weight (UHMW) polyester fibers.

The Japanese were quick to note the exceptional properties of aramid fiber when it was first developed in the U.S. and immediately sought to gain a proprietary position as a supplier of improved versions of the material. Japanese companies became involved as early as 1974, when DuPont was still issuing its initial patents. These early entrants included Tiejin Ltd., Asahi Chemical Industry Co., Toray, Toyobo, Ube Industries, and Unitika. The research involved many variations in composition or process technology. Tiejin was the most successful of these companies, producing HM-50, a novel diamine composition. This fiber has tenacity and modulus characteristics similar to DuPont's Kevlar, but with improved abrasion and impact resistance, as well as better hydrolytic stability—all at a competitive cost. The fiber has received high praise as a result of innovative synthetic polymer chemistry, but Teijin has not yet released it as a commercial product.

It is apparent that Japanese research on both current and future (e.g., HM–50) aramids will continue, particularly as pertinent U.S. patents expire in the next decade. Even while patents protect American firms from competition on their existing products, Japanese innovations such as HM–50 may soon command premium prices that can offset the costs entailed by lower production volumes.

Because of U.S. patent protection Japan has lagged behind the United States in aramid technology. Over the next decade, Japan is expected to catch up and be able to compete on an equal basis. Japan is similarly active in polyimide research. This material, which has a flexible aromatic ether linkage, gives good toughness and high elongation as well as high T/M. DuPont and Toray operate a joint venture which has been producing the material since mid-1985.

Other Japanese companies engaged in polyimide research include Ube Industries, Hitachi Chemical, Mitsubishi Chemical, and Mitsui-Toatsu Chemical. Ube has developed a low-cost synthesis of biphenyl dianhydride (BPDA) that has premium properties at prices competitive with American BPDA. Because of the stiffer all-ring molecular structure of the material, the film has higher T/M and adequate elongation, coupled with very high thermal resistance and improved hydrolytic resistance. It is not clear, however, whether these performance improvements will translate into additional sales in the near future.

In contrast to the patent protection that aramids will enjoy for a decade or more, most polyimide patents have already expired. Concurrently, the Japanese have exploited innovative research to develop several new compositions with improved properties. If current trends continue, it is likely that Japan will surpass the United States in polyimide research within the next few years and will seriously challenge American firms' market share early in the 1990s.

The Japanese are also within striking distance of the United States in the area of liquid crystalline polyesters, although actually surpassing American producers is likely to be considerably more difficult. This technology has been dominated in the recent past by Celanese through its Vectra resin, but Eastman Kodak, Sumitomo, and other Japanese firms have undertaken serious research. Sumitomo, through its licensing of the Carborundum technology, and Mitsubishi Chemical Industries, through its licensing of Kodak's XJ7 polyester resin, are particularly noteworthy. The Kodak resin has a low thermal resistance, but its high T/M potential and low manufacturing cost would make it attractive in applications not requiring extreme temperatures. Kodak's domestic test marketing failed, largely because of difficulties in polymerization and process technology in making a uniform homogeneous layer and in extruding thick objects with uniform properties. The Vectra resin, by comparison, has a higher melting point, is homogenous, and has a low melt viscosity due to its LC nature. It is still difficult to mold, but Japanese process engineering could become significant in this technology.

As incremental engineering improvements continue, Japan is likely to pull even with the United States. At that point, the American manufacturers will need to rely upon engineering and marketing expertise, along with a wide technology base in related composite development to stay competitive.

The final major class of high T/M materials is the group of UHMW polymers. These low-cost polyethylene materials exhibit the highest known specific tensile strength and hydrolytic stability, but they currently can be manufactured only through an expensive, inefficient process. The only two companies active in this area are the Dutch State Mines/Toyobo joint venture, based in Japan, and a Mitsui Petrochemical Industries project.

The DSM/Tobobo plant produces fifty tons per month, but the Mitsui facility can handle only three tons per month. The U.S. is at best parallel to the DSM/Toyoba basic research, but it is likely well behind some important technological breakthroughs Mitsui has accomplished to date.

Engineering Resins and Matrices for Composites

Engineering resins and composites are high-performance polymer-based materials capable of functionally and aesthetically replacing metals and other conventional materials. The principal advantage of engineering resins and composites lies in the ability to tailor property sets for specific applications. These properties include processibility, low density, mechanical characteristics (especially toughness), flame resistance, use temperatures, aesthetics, and cost.

Property control is achieved by the choice of chain chemistry, microstructure, and the number of included phases. In general, the mechanical properties and the use temperature will increase as the stiffness and/or aromaticity of the backbone chain is increased, often at the cost of processibility and aesthetics. Molecular chain orientation will markedly increase axial mechanical properties while introducing property anisotropy to the finished part. Multicomponent materials range from continuous fiber reinforced structural composites with specific mechanical properties in excess of metals, to polymer blends and alloys designed for improved toughness or processibility.

The key resins are polyamide (nylon), polyethylene terephthalate (PET), polybutylene terephthalate (PBT), polycarbonate (PC), and modified PPO or PPE (blends of polyphenylene oxide or ether with polystyrene polyacetal. The main composites are those used for fiber reinforcement: carbon, glass Kelvar, and aramid.

Japanese research in this area is characterized by a dearth of basic conceptual scientific papers or generic inventions and a flurry of product development activity, reflected in an extremely high number of patents issued to Japanese companies for engineering resin technologies. Between 1981 and 1985 Japanese companies received over 2000 patents (compared with 800 received by U.S. firms during the same period) for innovations on generally known but not optimized product and process developments.

Use of engineering resins is dominated by the electronics and transportation industries. Japanese companies who manufacture or market these materials are virtually all involved in business arrangements with American firms. Since the domestic demand in Japan is quite strong, U.S. engineering resin producers have benefitted from these joint ventures and marketing arrangements. In the future, especially after many of the key patents expire, Japanese firms will want to export these resins. The Japanese intention to enter foreign markets was reinforced by decisions to expand plant capacity.

Japanese resin technology is equivalent or superior to that of the rest of the world. The focus is on developing additives, fillers, and blends for specific applications rather than on inventing base resins. Japanese innovations result primarily from incremental engineering improvements and constant attention to detail. A key factor in Japan's success is the industrial group structure, which make it possible for automotive and electronics companies to provide resin manufacturers in their own group with timely feedback about new products and processes. Good lines of communication between resin producers and resin users in electronics and automobiles are much less common in the United States.

In matrix materials and composites, Japanese products and presence in the world marketplace lags the United States by two to three years. Lack of success is not due to technical factors, but to the absence of domestic aerospace and military markets and therefore the absence of an internal product development feedback loop. Japanese interest and presence in the matrix market is increasing, and should be aided by MITI-sponsored programs.

Polymers Used in Electronic Applications

Research in the use of polymers for electronic applications—for the fabrication and packaging of integrated circuits (ICs)—is split between efforts at understanding the chemistry of materials already used in such applications and efforts at developing new materials to address the problems and requirements of integrated circuit manufacturing today. These materials represent a critical element driving the electronics and computer industries. In the United States, most of the research is devoted to understanding existing materials; 75 percent of U.S. development in this field is performed by IBM or AT&T Bell Laboratories.

In Japan, by contrast, advanced development activity is diffused among a number of firms, including Hitachi, NEC, NTT, Japan Synthetic Rubber, Toya Soda Manufacturing, Sumitomo Chemical, and others. The Japanese industry benefits from greater competition and from its vertical integration, which better equips it to bring new innovations to market. If Sumitomo Chemical develops a new material, for example, it works with Sumitomo Electric or NEC to evaluate and produce the material. Such effective feedback channels are rare in the U.S. industry; only IBM and AT&T are sufficiently integrated to be able to develop prototypes and set up test and evaluation feedback loops internally. In an industry where the research-development manufacturing cycle has shortened from an average of eight years to somewhere between two and three years, the ability to exploit new technology and bring it to the market quickly will be the decisive factor in commercial success.

Three broad classes of polymers are used in electronic applications: lithographic, packaging, and encapsulant materials. The individual elements of an IC are defined by a lithographic process, in which the desired pattern is first defined in a resist layer that is usually a polymeric film spin coated onto the substrate. This pattern is then transferred into the substrate by etching, ion implantation, or diffusion. The materials commonly used in this process are available in commodity volumes in Japan as well as the United States.

As the minimum resolution required by IC designs exceeds the capabilities of the conventional photolithographic process, new materials and processes will be required. Researchers in both the United States and Japan have been developing new chemical lithographic techniques. Two generic types of techniques are currently being researched. Single-layer processes include deep-ultraviolet, electron beam, and X-ray technologies. As integrated circuits become smaller and more complex, the limits of these single-layer techniques begin to constrain further development. In particular, problems associated with substrate topography and surface reflectivity are overwhelming unless multilayer techniques are used.

The U.S. interest in both single-layer and multilayer processes is quite broad. Active research continues in single-layer technologies, including methacrylics (positive etching), acrylates (negative etching), sulfones, novolac/sulfone composites, vinylphenol esters, vinylphenol/azide composites, styrenes, and silyated resins. Of these, Japan has been the origin of only two, vinylphenol/azide composites and silyated resins. Although Japan has not originated the exploratory research, the Japanese already have commercially available products in these areas, whereas the United States has none. For example, novolac/sulfone composites and styrenes were both explored initially in the United States, and American researchers are still actively pursuing these chemical structures, but only Japan have marketed them in commercial quantities. U.S. research results in this area have remained inside the laboratories of IBM and AT&T, for proprietary use. This is a reflection of Japanese abilities in engineering and using known chemistry rather than in developing new chemistry for resist applications. It also reflects the determination of the Japanese industry to concentrate both on developing working processes for new materials and on finding markets for them.

Since the initial report of a silicon containing a two-level structure in 1981, many systems have been developed in both the United

States and Japan. The Japanese efforts seem concentrated at NEC, Hitachi, and NTT, while the U.S. research is concentrated at IBM and AT&T Bell Labs. General Electric (GE) has developed a significant advance in contrast enhancement materials, an area so far ignored by the Japanese. It is entirely possible, however, that as the American industry expands its basic research efforts with these materials, the Japanese will become interested and initiate some development research. To date, they have shown little interest in developing new lithographic materials, preferring instead to borrow U.S. technology.

In packaging materials, the Japanese seem more effective in advanced development work, using engineering skill and existing knowledge to solve targeted problems. The polymers used in packaging are similar to those used in other applications, enabling the Japanese to shift knowledge gained in one area to another without needing to regenerate all of the basic research. This capability will work increasingly to Japan's benefit as the demand continues to grow for polymeric coatings to insulate dielectrics and as passivation layers for semiconductors and thin-film hybrids.

The most promising materials for these applications are polyimides and cycled polybutadiene resins. These materials are soluble and thus are readily spin-coated onto a substrate. Consequently, the surface they create is relatively flat, which is more suitable for further manufacturing processes before the chip is ready for sale. The packaging materials also exhibit high temperature stability, insulate well, are unusually resilient, and can be obtained with high purity.

Despite the recognized future demand for polyimides, few companies in the United States seem to be exploring them. Since Japan is heavily involved in designing and marketing polymeric dielectric coatings, many American technology firms have turned to Japan for source materials. The only major American firms actively pursuing polyimides for integrated circuits are DuPont and IBM, which keep their results proprietary. The major Japanese participants are Toray, Ube Industries, Hitachi Chemical, and Japan Synthetic Rubber; in general, these companies are well ahead of U.S. efforts at every step along the development process, and are widening their lead.

The final process used in applications for the electronics industry is encapsulation, in which devices are coated with layers of extremely pure materials to prevent contamination and water penetration and to diffuse heat generated by the chip. Little tolerance exists for stress between the chip and the encapsulant, since high stress may lead to cracks and thus contamination. This stress is likely to occur because the chip and the encapsulant material will likely have different thermal expansion coefficients.

Japanese and American efforts in encapsulation seem roughly equivalent, although the Japanese are picking up the pace in advanced development and product engineering. In particular, Nitto Electric has developed a siloxane-spoxy copolymer that exhibits improved performance over existing materials, and epoxy novolacs are commercially manufactured only by Sumitomo Chemical and Nippon Kayaku. However, Japanese manufacturers tend to market such products before the processing technology is fully understood, while U.S. firms tend to be more cautious.

Membrane Separation

Membrane separation is the process by which a stream of liquid or gaseous material is passed through a membrane material which separates the desired permeate stream from the remaining nonpermeate stream. Much of the early innovation in membrane separation technology occurred in the United States in the 1960s, but in the ensuing two decades the Japanese have expressed and intensified their interest. While the United States still appears to have a substantial lead in this area, Japan's current level of activity suggests that it will become a major competitor within several years. The technology is particularly useful in chemical processing and in the small-scale production of highly valuable biotechnology products.

Although Japan has been interested in membrane separation technology since the early 1970s, a major commitment has been apparent only since 1980. In that year, MITI began funding and coordinating research on advanced hydrogen separation membranes and on basic technology issues in the field. By 1983, MITI was allocating $7.1 million annually to membrane research alone, the lion's share to basic technology for future industrial usage. The agency established a ten-year timetable for focused membrane research, with the initial period (1981 to 1984) targeted at clarifying the separation mechanism of membranes and screening candidate polymers. The second phase (1985 to 1987) focuses on developing new membrane materials and systems, on understanding and developing ultrathin membranes for high flux, and on designing new types of modules and membrane systems. The final phase is geared to fabricating high performance membrane systems.

MITI's agenda for new membrane materials bears a striking similarity to its Fifth Generation Computer System (FGCS) project, also a ten-year program divided into three phases. In both programs, the initial period is intended to understand the properties and attributes of current technologies. The middle phase is then pegged as a period of highly creative advanced development and significant process or material innovation. Specific elements of a total system are expected to be developed and mastered during this stage. In the final phase, MITI expects researchers to combine the results of all the individual projects into a holistic system that can be produced in large numbers.

In the Fifth Generation Computer System (FGCS) and the membrane programs, MITI has acted both directly through the government laboratories and indirectly through industrial companies to foster research. MITI directly coordinates work at the National Chemical Laboratory for Industry, the Industrial Product Research Institute, the Research Institute for Polymers and Textiles, and at nine industrial partners, including Mitsubishi Chemical, Sumitomo Electric, Toray, Toyobo, and Asahi Chemical. The facilities at the national labs are particularly advanced; no U.S. government labs seem to have comparable resources. Through its Industrial Structure Council, MITI follows the work conducted at fifteen other companies, including Fuji Photo Film, Matsushita Electric, Mitsubishi Rayon, and Ube Kosan Chemical.

Paralleling the MITI funding, the Ministry of Education and Culture has supported significant membrane research within the academic community. In 1983, for example, it spent $1.8 million on synthetic membrane materials and $2.2 million on biomembranes. This largesse has been spread over a large number of university beneficiaries and a wide variety of topics. The university work concentrates on finding new materials rather than on understanding the properties of existing materials. Other government agencies have also shown interest, but with far less funding. The Ministry of Agriculture, Forestry and Fisheries spent $400,000 in 1983 on membrane research in the food industry, and the STA appropriated $500,000 for exploratory research in the area.

Most of the MITI and private sector membrane research can be classified into four categories, determined in part by the degree of control (versus passivity) inherent in the process. The four categories, ranging from the most passive to the most clearly controlled by the chemical solutions used, are microfiltration/ultrafiltration, dialysis, reverse osmosis/pervaporation, and gas separation. The first three involve liquid separation and have been commercially available in Japan and the United States for ten to fifteen years; the Japanese products are currently believed to be roughly equivalent to U.S. membranes.

The gas separation field, by contrast, has become a commercially viable entity only within the past few years, and the U.S. industry has a sizable lead through the early efforts of Monsanto, Dow, W.R. Grace, and Separex. A major thrust of MITI's effort is aimed at gas separation, however, and as the market continues to evolve rapidly Japan may catch up in some areas.

Commercially, gas membrane techniques are used for hydrogen separation from nitrogen, methane, and carbon monoxide. Significant evolving markets are likely to emerge for nitrogen in blanketing valuable materials, for oxygen for medical and combustion applications, and for carbon dioxide (from methane) in enhanced oil recovery operations. The largest part of the Japanese effort is concentrated on hydrogen separation, with smaller efforts under way in nitrogen and oxygen separation. The first generation of membranes for hydrogen are reportedly close to entry into the market. These membranes are based on composite structures, in some instances using cross-linked selecting layers.

Of the three membranes closest to the market, Ube Kosan's appears to be the most significant. It is roughly equivalent to membranes marketed by Monsanto, but it has the advantages of higher selectivity and the ability to operate at higher temperatures. Even before this generation of membranes has reached the market, successive generations are also reportedly in the works at Japanese labs.

The Japanese work in oxygen and nitrogen separation has been less impressive. Concentrated in the industrial companies and not at the national labs, this work has resulted in materials with unusually high permeabilities but extremely low selectivities. Asahi Glass is close to marketing an oxygen separation system, but Japan otherwise seems unlikely to accomplish any significant milestones in the near future in separation of gases other than hydrogen.

In liquid separation, however, which in Japan is already a $300 million market, the Japanese are making impressive strides. Almost 75 percent of the market is accounted for by dialysis, while about

15 percent comes from microfiltration and ultrafiltration. Electro-dialysis and reverse osmosis together generated only $11 million in 1983. The commercial success of such relatively passive systems as micro- and ultrafiltration and of dialysis has spurred the Japanese to additional research in this area. A comparison of Japan's extensive research program with its relatively small domestic dialysis market suggests that Japan has its eye on export markets for these separation membranes. Potential applications for these membranes abound in medicine, agribusiness, and other field. In the dairy industry alone, ultrafiltration is used for concentrating whole milk and skim milk, fractionating whole milk or cheese whey, and separating inhibitors of microorganism growth for fermentation. In addition, cheese whey is concentrated using reverse osmosis membranes.

The primary motivation behind current research in these passive membranes is not the development of better materials, but rather is the prevention of surface fouling in currently available membranes. Because fouled membrane surfaces cause expensive shutdowns in the continuous flow processes typical of industries that rely on membrane separation techniques, a breakthrough in fouling prevention could provide a significant competitive advantage. In the absence of a revolutionary advance in prevention, which is unlikely, developments in pretreatment and module engineering will be necessary to control fouling and facilitate cleaning. Traditional Japanese engineering skills and attention to detail make it likely that they will be aggressive in making these incremental improvements. This research is driven in part by cost reduction, and therefore is likely to be undertaken by industrial rather than academic groups. This strengthens Japan's hand. The United States is better positioned, because of a stronger academic materials science field, to realize performance enhancement through better or more complex membrane surfaces.

Some research is also under way to develop new applications for existing membrane materials. This research is supported indirectly by government incentives to consumers and firms to purchase membranes for innovative uses. For example, the Ministry of Public Health and Welfare encourages tax-free purchases of membrane systems for their energy efficiency, and several recycling programs are using membranes to convert waste into useful materials in food processing, water reuse, and other areas.

While reverse osmosis is not as commonly used as dialysis or micro- and ultrafiltration, it is still a subject of important research.

In particular, this technique is used in the desalinization of brackish and sea waters for drinking and high-purity process applications. In principle, the technology is suitable for organic and aqueous systems, but in practice difficulties with membrane stability have deterred significant efforts. Perhaps because of these difficulties, the Japanese see desalinization applications as the primary targets of current research. In particular, efforts are under way to develop the more robust membranes needed in the Middle East, where seawater is warmer and saltier than elsewhere.

Over the longer term, the Japanese research community appears interested in dealing with a much broader range of reverse osmosis applications than is currently considered possible. Membranes capable of separating organic molecules from water or from complex organic solutions are being developed, but the required membrane materials are not yet available. These membranes are likely to be composite materials, such as the crosslinked FT-30 currently available in the U.S. MITI has recognized this deficiency and is apparently funding considerable efforts in composite membrane development.

Biopolymers

The widespread use of polymeric materials in medical and biological applications has mushroomed in the past two decades. Such materials are currently used for artificial skin, pacemakers, heart and vascular prostheses, intraocular lenses, sutures, prosthetic devices, and a host of other applications. The materials used for these applications—particularly those used in connection with human bodies—must meet stringent conditions. For example, the biopolymers used for hip replacements must remain strong and rigid, while heart valves and arterial grafts must retain their pliability for periods of over thirty years. Moreover, these materials must not cause any incompatibility with the host's blood or tissues. Even small imperfections or impurities in a biopolymer can cause tissue inflammation, edema, and antigenic reactions.

Since their first commercial use in the 1950s, biopolymers have become a big business. Early applications included contact lenses, dental composites, and prosthetics; through the 1960s, polymeric drugs and drug carriers became common, and in the 1970s time-release polymers were marketed. Japan has recognized the enormous commercial potential inherent in biopolymers and begun a significant research effort, but it is still well behind the efforts underway at American pharmaceutical companies.

Nevertheless, the Japanese industry has progressed steadily. In 1980, Japan was considered five years behind the United States; by 1985, the nation was only about one or two years behind. About 10 percent of Japanese technology is already equal to or better than its U.S. counterpart, about 70 percent is less than two years behind, and 20 percent is still more than two years back.

Moreover, Japan is likely to continue its rapid pace, primarily through redirected funds rather than from newly appropriated money. Chemical companies as well as firms in other sectors are now becoming involved in biopolymer research and development. Toray, for example, had in 1985 about 200 professionals in the area, with plans to transfer about 40 to 50 additional workers from other areas within the next decade. Asahi Glass had only four professionals assigned to biocompatible materials, but it expected to increase that level to twenty personnel with comparable funding within the next five to ten years.

In addition, several government laboratories are supporting biopolymer research, particularly at the National Chemical Laboratory for Industry, the Fermentation Research Institute, and the Research Institute for Polymers and Textiles. Much of the national laboratory and industrial research is funded through AIST, which is providing over $7 million annually through an eight-year program. The Japanese strength in biopolymers lies in the ability to produce high-quality products innovatively from existing technology. Japan has a clear commercial advantage in that its government has actively supported biopolymer research and has refrained from extensive regulation and monitoring of animal testing and human evaluation.

The Japanese industry excels at developing commercial applications for a material once it has been discovered; its weakness is an inability to develop new polymer systems. One reason for this is the poor test and evaluation instrumentation found in most Japanese universities and even in some industrial laboratories. These problems have kept Japan from making much progress in some areas where MITI has sponsored research, such as time-release polymers, artificial organs, and biomaterials. In immobilized agents, however, Japanese work is far ahead of that in the United States. In this area the Japanese not only have proven manufacturing systems, but the basic and applied research to support them.

SUMMARY

Japanese research in advanced polymers is summarized in Figure 5-1. The national commitment to the field is evident in the fact that in almost all areas of basic and applied research, including those in which Japan is behind the United States, the Japanese industry is gaining ground.

The polymer industry evolved quite differently in Japan and the United States. The U.S. field came into existence because of a sustained funding commitment by the Department of Defense. Applied research in corporations was balanced by conceptual research at national laboratories and universities which focused on understanding rather than developing materials. U.S. chemical companies have grown larger through acquisitions and mergers.

The Japanese industry, however, relied heavily on U.S. licenses rather than on indigenous basic or applied research programs. Attention to detail in manufacturing and incremental process and engineering improvements has been the key to commercial success. The Japanese chemical companies have tended to spin off polymer subsidiaries rather than to pursue centralized growth. The areas in which Japan is behind the United States tend to be the ones where the product feedback loop has been difficult to establish because Japan lacks aerospace and military markets.

Finally, the Japanese industry has been able to establish and stay with medium to long term goals. This commitment has called for sustained investment in research even during difficult economic periods and a sustained effort to develop new processes for existing polymers even when commercial success did not materialize immediately.

Figure 5-1. Overview of Japanese Research in Advanced Polymers.

CATEGORY	BASIC RESEARCH	ADVANCED DEVELOPMENT	PRODUCT IMPLEMENTATION
HIGH STRENGTH/ MODULUS POLYMERS	–↗	–↗	–↗
ENGINEERING PLASTIC & MATRICES	–↗	0↗	+↑
POLYMERS FOR ELECTRONIC APPLICATIONS	0→	0↗	0↗
MEMBRANE SEPARATION TECHNOLOGY	0↗	0↗	0↗
BIOPOLYMERS	–↗	–↗	–↗

CODING SYSTEM — JAPAN COMPARED TO USA:

PRESENT STATUS

+ AHEAD
0 EVEN
– BEHIND

RATE OF CHANGE

↗ GAINING GROUND
→ HOLDING CONSTANT
↘ LOSING GROUND

6 MECHATRONICS

The Japanese define the field of mechatronics as a multidisciplinary integrated approach to product and manufacturing system design. This approach encompasses the next generation of machines, robots, and smart mechanisms necessary for performing work in a variety of manufacturing and non-manufacturing environments. Software and related ancillary technologies are also considered part of mechatronics. Thus, mechatronics combines the disciplines of electrical engineering, mechanical engineering, and computer science.

In the United States, these activities have been aggregated loosely under the rubric of "factory of the future" or simply factory automation. The techniques are concerned largely with the integration of the product design with the manufacturing system. They do not attempt to integrate or coordinate the manufacturer with its supppliers.

The difference between U.S. and Japanese approaches to the field reflect differences in institutional and industrial environments. The Japanese concept of mechatronics represents a new way to integrate product design, manufacturing, and supplier companies. Japan considers the supplier an integral element of the manufacturing process rather than merely a provider of equipment or raw materials under specified terms. Japanese companies have large production technology centers which design and produce the necessary systems; the user and the supplier are different divisions of the same company.

Feedback loops between producers and users enable these companies to use advanced design and manufacturing techniques to develop new products and build them independently of the external market.

In the United States only a few vertically integrated companies have the ability to act as their own suppliers; General Electric and IBM are two examples. Vertical integration is increasingly being replaced by multiple suppliers that are often dispersed widely across the United States or around the world. Even IBM has shifted toward multiple suppliers in its newer products; its Personal Computer (PC) family is built almost entirely from parts acquired from dozens of different suppliers.

Thus, U.S. manufacturers are often constrained by suppliers who are concerned about their own profitability and who will not embark on major investments without assurance of market acceptance and a corresponding return on their investment. Japanese companies have little trouble coordinating the basic relationship between the manufacturer and the supplier. The Japanese industry therefore benefits from a more aggressive, experimental, and flexible approach to the adoption of modern manufacturing techniques.

Because the fundamental notion of mechatronics is that the entire factory is an indivisible organism, the market for mechatronics systems is limited to companies that have large manufacturing facilities which handle many different products and components. If a plant handled only one component or made only one product, the flexibility that is the main selling point of Flexible Manufacturing Systems (FMS) would be irrelevant.

As a result, only companies with the largest and most capital intensive manufacturing facilities are likely to be interested in mechatronics. Yet these companies—particularly in the United States—are generally most risk-averse and do not want their plants to rely excessively upon the performance of any single outside party. American manufacturers have shown considerable resistance to the notion of any single company developing a large-scale mechatronics system within a factory. At the same time, many of these companies want (and need) to enhance their manufacturing facilities. They are anxious to move toward mechatronics systems, but in an incremental, gradual way; few want to take a single giant leap.

General Motors (GM) came to the rescue of many firms facing this dilemma when it purchased Electronic Data Systems (EDS), a

Dallas-based software firm. GM, in conjunction with EDS, has developed the Manufacturing Automation Protocol (MAP), which decrees a set of standards to which all vendors must adhere in order to sell manufacturing or information systems to GM. The standards are highly specific and pertain to data communications, factory floor devices, and several other areas. Because GM is one of the largest corporate purchasers in the United States, its standards carry enormous weight in industry. Many other manufacturers now specify MAP compatibility in their purchasing requisitions. The advantage of MAP to GM and its peers is that the protocol enables them to develop highly advanced manufacturing systems similar to Japanese mechatronic systems.

By removing the single-vendor obstacle, MAP is similar in effect to the vertical integration of large Japanese companies. In fact, what existed in the Japanese industry due to agglomeration and integration, American companies have accomplished by imposing standardized protocols.

The Japanese have gone a step farther than the United States in facilitating the growth of mechatronics through MITI's Industrial Standards Committee. Established in 1949, this group publishes national standards for industry; it already has approved six sets of standards that deal specifically with industrial robots, for example. The Japanese government encourages companies to cooperate to form these product standards. This stands in contrast to the United States, where collusion among companies is a violation of antitrust statutes.

Mechatronics is an evolving concept. The level achieved by most Japanese companies to date is the use of integrated teams of product designers, manufacturing, purchasing, and sales personnel acting in concert to design both the product and the manufacturing system with minimal technical complexity. For the future, however, mechatronics promises to be a means for implementing more advanced processes and systems technology. The current status of mechatronics can be addressed by examining two distinct levels of technology: systems and components. Within the realm of mechatronics systems, relevant technology sectors include: flexible manufacturing systems (FMS), assembly and inspection systems, vision systems, non-vision-sensing systems, manipulators and actuators, precision mechanisms, and software.

JAPANESE ORGANIZATIONS INVOLVED IN MECHATRONICS RESEARCH

Mechatronics Research in Japanese Industry

Because mechatronics includes primarily manufacturing, assembly, and related processes, the vast majority of research, both in the United States and Japan, has been conducted in corporate research departments rather than universities. MITI has provided some assistance, but this has consisted of support for programs in specific areas, such as flexible machine systems, rather than for mechatronics as a whole. Such programs will be discussed below in greater detail. The FGCS and the Language Translation projects discussed in earlier chapters have also had direct applications to mechatronics systems.

While the brunt of the research burden has fallen on private enterprise, the cost of research and development in mechatronics has proved to be enormous. Thus there is a strong incentive for cooperative research ventures in this field.

Mechatronics-Related Research at Japanese Universities

Japanese and American universities have both made significant contributions to mechatronics from within existing departments, although there have been very few, if any, large projects which encompass the integrating concept which ties the components of mechatronics systems together. It is extremely difficult for American universities to engage in interdisciplinary research because of the artificial boundaries between many university departments. The university reward structure in the United States generally discourages young, untenured and creative faculty from pursuing such interdisciplinary group projects, since these people must show clearly individual successes if they are to win tenure. Academicians who participate in large joint projects may suffer, since their interest is often apart from the mainstream of traditional disciplines and since the large projects hide the contribution of individuals in favor of the group authorship. The National Science Foundation has been trying to encourage interdisciplinary work through the establishment of Engineering Research Centers at leading science universities but the results are not in yet in how successful they have been.

The Japanese university system, for different reasons, also makes it difficult for faculty to pursue large interdisciplinary projects. The insularity of university departments in Japan results not from the need for individually authored publications, but from the funding system which allocates money only to established chairs. New projects can only be funded if new chairs are established. Since these chairs tend to be extremely small, however, no individual chair can encompass the breadth of scientific talent or afford the capital investment required for mechatronics research.

Japanese universities have made important contributions in research on actuators and mechatronics-related software, but there are no university-based programs for studying the science of mechatronics more broadly.

EVALUATION OF JAPANESE MECHATRONICS RESEARCH

The remainder of this chapter discusses the current status of Japanese research and development in each of the mechatronics technology sectors and assesses the prospects for future development in these areas. Where appropriate, these research efforts are compared to the U.S. approach, but the fundamental differences in industrial infrastructure make such comparisons difficult.

Flexible Manufacturing Systems

A flexible manufacturing system (FMS) can be broadly defined as any multimachine complex system that has substantial automated materials handling capabilities combined with some flexibility with regard to the actual operations performed. Often, the goal is a system that can produce a lot size of one unit as efficiently, inexpensively, and quickly. Japanese companies have been installing FMS systems for over a decade, and the market for FMS continues to grow rapidly.

In 1981, Yamazaki opened a new plant with two fully automated FMS systems. It has subsequently installed an additional two systems. Toyoda Machine Works, SNK, Niigatta, Makino, and other Japanese machine tool companies have also installed FMS. Toyo Kogyo, the maker of Mazda automobiles, has installed an impressive FMS facility in its fully automated facility. Parts are received, inspected, machined to precision requirements, and assembled in the plant, all with no human intervention at any point in the process.

Japanese companies have installed FMS with available and proven technology rather than waiting for more advanced systems. Those firms that have installed FMS have continued to design process improvements internally and introduce new hardware incrementally, with minimal disruptions to normal operations or efficiency. Consequently, the Japanese have become experienced users of FMS.

The development of FMS technology is best understood within the context of Japanese production engineering, which pushes the responsibility for potentially important decisions far down into the organization. Four levels of production engineering exist in a typical Japanese plant where FMS is likely to be used. The lowest level is the factory floor, where workers are encouraged to provide suggestions on ways to lower costs and increase productivity. Foremen on production lines can approve reasonable expenditures to modify the process. A second level consists of entry level engineers assigned to foremen. These engineers are trained in problem solving, and spend three to five years in an apprenticeship assignment. They, too, are authorized to make modest changes. The third level is the production engineer, who has gone through an apprenticeship and understands the technology well enough to oversee process improvements. These engineers meet directly with vendors of equipment and are responsible for designing new production processes. Finally, the advanced manufacturing research group is dedicated entirely to developing new processes. Not all Japanese companies have a group at this level, but many of the breakthroughs in FMS have had their genesis in such a group.

The key feature of the Japanese approach is that both the user and the producer perform R&D. In Japan, in fact, it is the research conducted by the user that drives the producer, whereas in the United States it tends to be the reverse. In addition, the continuous attention to the evolution of the manufacturing process by workers at all four of these levels has enabled the Japanese FMS industry to stay at the edge of technology and retain a leading position. It is possible that the experience of the steel industry, which Japan built with the most modern and cost-effective equipment, served as a negative example. Japanese steelmakers, once they had developed a large market share, failed to update their production technology; over time, they have lost the industry to Korea and Taiwan.

The primary government support in FMS has been through MITI's Flexible Manufacturing System Complex, sponsored since 1977 at Tsukuba and commonly known as "FMS with Laser." This national effort has two goals: first, to develop new processes, such as finding applications for lasers in manufacturing; and second, to develop flexible machine tool components that can be mated to create different types of machine tools. The MITI project and other Japanese FMS systems rely heavily on the development of automated CNC machine tools, materials handling systems, and robots. Japan is commonly credited with an enormous lead in the number of installed working robots, but this lead is largely a reflection of the way Japanese and U.S. industry define their terms. For example, Japanese consider simple manipulators to be robots, whereas U.S. industry does not. When these semantic differences are resolved, Japan is actually only slightly ahead of the United States in the use of robotics for manufacturing purposes.

Japan does appear to have a lead in research and development of automated materials handling systems. Researchers there are currently exploring the notion of Flexible Transfer Units (FTUs), which allow flexibility for design changes and new parts. Such a system is particularly applicable to the appliance, automobile, and tractor industries, where model changes from year to year often involve minor cosmetic modifications. These changes require new capabilities and machine tool components to allow the materials to be handled and machined properly. The FTU is a means of adapting to these changes with minimal disruption. The idea has already been put into practice at Nippondenso, which is using an FTU system to manufacture its line of automatic radiators.

Finally, FMS depends on computer-aided design and computer-aided manufacturing (CAD/CAM). It is through the use of CAD/CAM systems that products are designed and that the manufacturing process is integrated with the design process. While CAD/CAM is growing in Japan, it is still well behind the level achieved in the United States and Western Europe. The Japanese tend to buy U.S. CAD and production control systems and then push toward standardized product designs. Even in the United States, companies such as Intergraph, Computervision, and Calma (a unit of General Electric) that have pushed CAD technology forward have not succeeded in integrating CAD designs with CAM production processes.

Assembly and Inspection Systems

A mechatronic assembly system is an automated way of integrating components into a finished product and assuring that the product meets certain specifications without the use of sensing or feedback systems. Not all tasks are suited to mechatronic assembly, and the Japanese have excelled in recent years in determining which assembly processes are most amenable to automation and which require human intervention. For example, people perform tasks like attaching and mating fine wires, whereas if the wire laying is highly structured (as it is when attaching leads to a silicon chip), the process is automated.

Each company seems to have developed its own system architecture. Three general archetypes have emerged. Sony and other companies use a modular system, Hitachi and others use distributed systems, and Seiko uses a continuous flow architecture. There appears to be little interest in assessing whether any given system might be optimal for any specific types of products or components.

Instead, Sony and Hitachi are using totally dissimilar architectures to assemble the same basic videocassette recorder (VCR) mechanism. Both systems have highly integrated parts supply and product handling systems. Hitachi uses a distributed system composed of people, robots, and fixed automation. This approach, somewhat older than Sony's, relies more heavily on the fixed automation aspect of traditional assembly systems. The Sony system uses groups of proprietary modules and people, and is a direct outgrowth of its product design. The Sony assembly system consists of an x-y table, a frame from which the machine tools are hung, pallets for parts, and a pallet transporter. All of this is controlled by a dedicated Intel 8080 chip.

The continuous flow architecture is typified by Seiko's Hoop Integrating Assembly System. With this system, a finished product rolls off the line every 1.2 seconds, although the actual process for any given unit—its throughput time—is considerably longer. To accomplish this, Seiko uses totally new part feeding mechanisms with 99.9 percent reliability. Small parts are captured in reeled plastic strips, while larger parts are captured on short-braid strips of about twenty-four pieces. The advantage of the process is that Seiko can switch an assembly line from one product to another within ten minutes.

In Japan, vertical integration is common in automated assembly systems. The company that produces a certain product also designs and implements the process by which that product is assembled. Sony, for example, makes both the VCR and the automated assembly system used in producing the VCR. This vertical integration helps explain why so many different architectures for automated assembly exist in Japan. In the United States, only GE, IBM, and AT&T have attempted large-scale efforts at designing their own assembly equipment. Other firms may be capable, but the market dynamics suggest that it is more cost-efficient to buy the systems outside. Often, the outside suppliers chosen by U.S. firms are Japanese companies, such as Sony. After five years of developing and testing its FX assembly system, Sony began selling it to outsiders in late 1984. Other companies have followed suit. Consequently, many Japanese robots are licensed in the United States, while few U.S. firms are developing their own robots.

Japanese systems are optimized for their makers' own products, not for American products. Consequently, Japanese assembly systems cost the American firms less than internally developed system, but they are also less complex, capable, and flexible.

Above and beyond the integrating technologies discussed above, the Japanese mechatronics effort hinges on the successful development of specific components. These include vision and nonvision sensors, software, manipulators, and precision mechanisms, which can be combined in various ways to develop fully configured FMS, CAD/CAM, or assembly systems. The following sections discuss each of these building-block technologies.

Vision Systems

Vision systems enable a robot to sense when a task is complete or to determine which element of the task needs additional attention. The Japanese have been conducting important vision system development projects for nearly fifteen years, beginning with Hitachi's printed circuit (PC) board inspections in 1973. Initial applications used simple binary (black-white) vision, but more sophisticated systems have emerged in recent years. Japanese firms have already developed dedicated special-purpose hardware for computer vision. They are now working on very large-scale integration (VLSI) architectures for gray scale vision. As is the case at the systems level of mechatronics, the important research in this basic building-block technology is conducted almost entirely within the industrial setting. Hitachi (at its Production Engineering Research and Central Research labs), NEC, Toshiba, Fujitsu, Mitsubishi, and Komatsu have been leaders. MITI's Electrotechnical Lab and Kyoto and Osaka Universities have also conducted research in the field, but these programs seem quite separate from the industrial projects.

The development of sophisticated vision systems is tied closely to the consumer electronics field, where many of these systems are used. Japan has used its strong base in consumer electronics to develop a lead in vision systems. Sony, Hitachi, and Matsushita have all marketed low-cost solid-state cameras. The prominence of the industrial giants in the list of Japan's vision technology leaders contrasts with the U.S. industry, where the leading vision technology firms are small start-up companies. Giants such as General Motors and General Electric do have credible research laboratories, but they have frequently had problems converting their research into production technologies. The relative prominence of giants in Japan is partly due to differences in approach between the two industries. Japanese vision technology is geared toward commercial, low-cost, high-volume applications, while U.S. research is more often conducted with high-quality, price-insensitive military applications in mind.

Several important applications of vision systems demonstrate why electronics giants in Japan are interested in the technology. One of the first important applications was in lead bonding of transistor chips, developed by Hitachi in 1974. Binary vision systems are used in mask and reticle inspection in the manufacture of integrated circuits (ICs). Masks are critical in IC manufacturing, since they act as masters, and faults in the masks are replicated in the product. By inspecting and verifying masks in low volume with an accurate system, the high-volume inspection of manufactured circuits can be avoided. This process is used in about 50 percent of IC production to date.

An obvious extension of IC inspection is PC board inspection. This application has received significant attention at Fujitsu, Hitachi, and Toshiba. Because there are many forms of defects possible on PC boards, no existing system appears capable of doing the total inspection job. Moreover, because the inspection must be performed on the actual product—not on a mask—the expected volume of such an application is extremely high, requiring significantly more computational horsepower. Similar vision systems for inspecting solder joints, semiconductor wafers, and diode pellets have all attracted some attention, but these seem to be of less interest to Japanese companies at the moment.

Outside of the electronics industry, additional applications have emerged for vision systems. The electronic aspects of optical character recognition, which enables machines to read printed characters or symbols, are discussed in Chapter 3. The optical aspects have also attracted attention, both in the United States and in Japan. In the United States, Cognex and GE have specialized in reading markings on products and instruments for industrial applications, while in Japan, firms have concentrated on reading addresses, a special problem because of the Kanji characters.

Komatsu, the large heavy equipment manufacturer, is developing visually guided automated vehicles that could adapt to changing warehouse configurations and navigate around obstacles. Komatsu already has marketed automated vehicles that follow wires in the floor. The use of such vehicles can be found in many manufacturing companies, as well as in some unexpected environments. One of the earliest installations of automated vehicle systems in the United States was the robot mail delivery system at the U.S. Department of Commerce building in Washington, a more recent installation is a stocking system at the University of Michigan hospital in Ann Arbor.

The major requirement for these vision systems is high-performance computation. Even modest systems are computationally expensive, about two orders of magnitude beyond the capabilities of the most powerful current microprocessors (e.g., the Intel 80286 used in IBM's Personal Computer AT, or the Motorola MC68000). Hitachi and Fujitsu have developed special-purpose logic for inspection systems based on simple binary vision, roughly equivalent to the products offered in the United States by Comtal, Vicom, and Westinghouse Auto-Q. Hitachi has gone beyond this level with two new chips, one of which is currently available to Japanese research laboratories. The chip contains four processors and can process a 256 bit by 256 bit image in 11 milliseconds. The closest U.S. chip seems to be the Radius chip made by Hughes Aircraft, but Hughes is not a commercial IC manufacturer.

The future for vision systems research lies in the realm of VLSI, which can generate the computational speed necessary for high volume applications. Hitachi, NEC, and other firms all have major projects in this area, while U.S. firms have not yet begun significant VLSI vision system research.

Non-Vision-Sensing Systems

Over 300 Japanese companies are active in producing specialized non-vision sensors. Much of this activity is directly tied to the robotics industry, a major consumer of sensors. The Japanese classify all sensors as either internal or external. The internal sensors are those which concern only the robot itself: how its manipulators are configured, how the temperature may be creating thermal expansion in the manipulator arm, etc. The external sensors allow the robot to interact with its task.

Internal sensors are used in robots to detect position, angle, force, torque, and pressure. External sensors are used in navigation, speech, touch, and information processing. The position and angle sensors operate by determining the internal characteristics of the robot, rather than by measuring against some external landmark. Knowledge of the robot's motor speed, the gearing details, and the arm geometry enable the sensor to calculate the arm's exact position from the robot's pattern of movement. Matsushita has recently achieved a price/performance breakthrough with its development of small, high-precision cartesian assembly robots. These units use precision helical ball screws to provide accurate motion at costs of about $10,000 per actuator. Each actuator is accurate to within 25 microns. The Japanese are also uncontested leaders in high resolution digital magnetic position measuring systems. Sony's Magnescale, for example, has no U.S. counterpart.

In other types of position-sensing devices, however, U.S. research compares favorably with Japanese work. Few manufacturers in either country have developed useful ways of monitoring force and torque accurately. IBM has begun incorporating this capability into its production robots through silicon strain gauges at the robot's wrist, and Kobe Steel and Kawasaki use deburring robots employing force feedback information. Fujitsu has also developed an active compliance device for precision assembly. Modest applications of similar technology can lead to useful capabilities. One example is the simultaneous measurement of both the torque and the normal force exerted by a drilling robot, which provides information on the sharpness of the drill bit. A related area, pressure sensing, is well-advanced both in Japan and the United States. This is a highly developed technique that uses diffused silicon gauges fabricated on a thin silicon pressure-deformable diaphragm.

As sensors for providing a robot with information about its own position and its immediate environment are perfected, new applications have emerged to enable the robot to interact with the environment. Ultrasonic sensors can be used as navigation aids for mobile robots. Japanese researchers at Tsukuba have extended this concept to the development of a prosthetic guiding robot for the blind. Both Polaroid in the United States and the Hilare organization in France are also working on ultrasonics. The biggest area for ultrasonics now is in medical imaging, but potential applications exist in detecting defects in nuclear reactor pressure vessels and in monitoring weld integrity. Mobile robots are a convenient way to make such measurements on large structures.

The need for mobile robots and larger autonomously guided vehicles is growing. These larger "smart carts" are becoming important for material transport in factory and warehouse environments, and are beginning to be used in semiconductor VLSI clean room facilities. At present, commercial technology is limited to carts that can follow a cable or a stripe of paint. By eliminating the need to follow a predetermined path, these carts can cover a much broader range, but doing so requires a navigational sensing capability.

One promising possibility for navigational sensing is a combination of inertial (i.e., gyroscopic) guidance and odometry (i.e., measuring distance as a function of wheel rotations). The key obstacle to such a system is the lack of a low-cost, low-drift gyroscope. Superb ring laser gyroscopes have been developed in the U.S. for military applications, but so far, these are much too expensive for factories, and little U.S. research is targeted at reducing their cost. In Japan, Honda has produced a low-cost gas jet inertial guidance system for automobile navigation called a "gyrocator" which seems well suited for factory smart carts.

As sensors for factory uses become more advanced, Japan hopes to develop sensors for uses in more hazardous environments where human backup is not possible. MITI has sponsored the Jupiter Project, intended to develop robots for use in hazardous situations. Potential results of this program include robots for radiation environments that employ scintillation counters or ion chambers adapted to the specific robotic use. Fire-fighting robots will need smoke detectors, thermometers, optical pyrometers, and chemical vapor sensors. These robots may be operated either by remote control (teleoperation) or autonomously. While teleoperation is preferable, some applications will require an autonomous robot. Japan has made considerable progress in this area.

Manipulators and Actuators

A manipulator is a chain of links, joints, actuators, and sensors governed by an attached controller. Because one end of this chain is fixed in place, workpieces or tools affixed to the open end can be moved and rotated in a predetermined way to conduct the actual manufacturing process. The actuator is the motor that moves one link relative to an adjacent link at their common joint.

Both university and industrial labs in Japan and the United States are conducting research on manipulators and their constituent parts. Universities and industry focus on different issues. Industrial labs have targeted pragmatic issues, such as increasing speed, reducing weight, miniaturizing systems, and instituting computer control. These design objectives are directly related to the physical properties inherent in moving a mass through space, and tie directly to cost, durability, and performance. In the universities, research is still concentrated on narrowly defined theoretical problems, such as sensory perception, positional accuracy, modularization, and simplification.

While most manipulators are attached to computer-controlled robots, not all manipulators are computer-controlled, nor even remotely capable of sensing anything about themselves or their surroundings. Two-thirds of the manipulators being produced today are still of the fixed sequence class, which means that the sequence of motions and operations they perform cannot be changed easily. Of the total, perhaps less than 1 percent are intelligent, employing active sensing of themselves and their task and using that sensing to modify their movements and operations.

Although few current Japanese manipulators are intelligent, Japan still defines the state of the art in manipulator technology. Fujitsu's MicroArm, which was developed to assemble optical fibers, operates through six degrees of freedom with a precision of 1 micron; typical accuracy for commercial models with similar degrees of freedom is 50 microns. Each degree of freedom indicates an independently controllable motion. Equally impressive is the Selective Compliance Assembly Robot Arm (SCARA), a manipulator design with four degrees of freedom, developed at Yamanashi University in 1978. This type of robot is available from many different suppliers and has been gaining popularity both in Japan and in the United States.

Another major new type of manipulator is the direct drive arm being developed in the United States at Carnegie-Mellon University, MIT, and Adept, Inc., and in Japan at MITI's ETL, Mitsubishi, and Yokogawa-Hokushin. The actuators for this type of manipulator are low-speed, high-torque electric motors that are now practical in a variety of designs. This type of motor eliminates the problems that have historically stemmed from the use reduction gearing, since such gearing is no longer needed. The direct drive arms are also noteworthy because of the significant speed and acceleration improvements they provide.

Multijointed manipulators of seven or more degrees of freedom are being developed for inspection tasks in difficult spaces, such as nuclear reactors and jet engines. A variable-length manipulator is being developed at Tokyo Institute of Technology; it can enter a small opening and surmount an obstacle to arrive at a goal position. These and other manipulators are being developed with teleoperated control systems. While such controls can be clumsy, they retain human control and provide a convenient transition between direct human manipulation and complete automation of the manipulator.

In order for these manipulators to work effectively, their component parts must also be effective. Thus, research on the links, joints, and actuators required for manipulators has been ongoing for many years both in Japan and in the United States. The manipulator's components are closely analogous to the human arm. The links, like bones, provide the structural integrity and move the end-effector, or hand, to its goal position; the joints, as in the human arm, allow one link to move in a direction separate from its adjacent link; and the actuators, like muscles, convert energy into mechanical motion by exerting the force needed to move the links to their intended position.

A major link technology development is the use of carbon filament reinforced plastic material, which is lighter and stiffer than previous materials such as aluminum. Hitachi, Mitsubishi, and Shin-meiwa are exploiting such materials. Joint research is concentrated on providing greater manipulator flexibility. By using a two- or three-dimensional joint, a link can be moved with much more freedom relative to its adjacent link than if the traditional one-dimensional joint is used. Toshiba uses a joint with two degrees of freedom in its multijointed manipulator, and Tokyo Institute of Technology has joints capable of three degrees of freedom.

Significant progress has likewise been reported on a wide variety of actuator motors in Japan. Direct drive motors with impressive performance characteristics have been developed by Hitachi, Sumitomo, Hokushin Electric, Fanuc, and Toshiba. Hitachi and Chuo University have improved fluid motors (i.e., pneumatic and hydraulic), and have used improved digital control technology to yield better real-time valve control. Thermal motors are also under investigation, but they are not as close to use in commercial manipulators yet. Tokyo Institute of Technology, Tokyo Electric College, and Hitachi are all experimenting with nitinol wires in thermal motors, since that material has a thermomechanical operating time of about one second. Hitachi has investigated the use of nitinol wires as thin as 2.5 microns in order to achieve faster operating times with such systems. Finally, Funitsu has conducted extensive research on the gearing of actuator motors. The gearing must both reduce the output of the motor to a level useful in the actual movement of links, and retain accuracy over time. Fujitsu has developed improved harmonic drive devices that provide significant size and weight advantages and that simultaneously retain their accuracy over extended periods. Toshiba and others have begun to follow similar approaches. Many other manipulator manufacturers are employing a variety of harmonic drives.

Precision Mechanisms

These electromechanical devices are used in mechatronic systems for high precision robots, semiconductor fabrication, and computer peripherals. The term "precision" refers to measures of accuracy, resolution, and repeatability. The development of precise, intelligent electromechanical systems lies at the core of many emerging technologies. For example, high-precision robots can be used in demanding assembly tasks such as the assembly of Winchester (rigid) disk drives, optical disk drives, videotape transports, optical communication elements, hybrid semiconductor circuits, and a wide variety of small electromechanical systems.

In the mid-1980s, Japan was the only source for high-precision robots in the world; Fujitsu's machines could achieve 5 micron accuracy and 1 micron repeatability, for example, compared to the 25 micron capability demonstrated by the most precise U.S. robots. Likewise, the aforementioned SCARA robot arm was a uniquely Japanese accomplishment; several U.S. manufacturers have since licensed the technology. The primary distinguishing attribute of this arm, aside from its exceptionally high accuracy, is its unusual rigidity in one dimension (the dimension in which force is being applied) and pliability in the two perpendicular directions.

Despite the current Japanese lead in high-precision instruments, much of the important development work was performed in the United States. The Remote Center Compliance device, investigated by Fujitsu for the production of Winchester disk drives, has its origins in work at Draper Laboratories as early as 1972. IBM has reported work on conceptually similar devices in the past two years as well. Related work on a much larger scale has been reported by the Carnegie Mellon Robotics Institute. This work involved the construction of a computer controlled remote center compliance device, but it does not have the high-speed two-dimensional feedback characteristic of the Fujitsu system.

The excellent performance of the Japanese robots is not primarily attributable to superior position sensors and actuators, although these components are essentially comparable to U.S. components. Rather, the Japanese excellence derives from the painstaking and original mechanical design and construction of the robots. These devices are usually constructed by a group that is closely connected to the developers of the product the robot will assemble. This close integration has a direct and beneficial bearing on the design process; the product designers can optimize the product to the manufacturing process, and the robot designers can optimize the manufacturing tools to the product.

Software

None of the FMS hardware or inspection systems can work without software to guide their operations. A machine tool needs a separate program for each component it handles, and an FMS system must have an overarching software program to coordinate all of the individual CNC and materials handling units. Japan's progress in mechatronics software closely mirrors its progress in computer software. As Chapter 3 describes, Japan has concentrated almost exclusively on hardware-related research and is consequently far behind the United States in software. In mechatronics, too, software has received considerably less attention than hardware.

The hardware emphasis is clearly evident in the FMS with Laser project. This program makes virtually no effort to coordinate or manage the development of software. In fact, it has been reported that the Ministry of Finance was concerned over the amount of money spent on software development in previous projects and did not allow the FMS project to include any software in its budget. Because software is essential to the completion of an operational system, each company on the project developed its own software and padded its hardware budget to cover the software development costs.

The effects of this neglect detracted significantly from the quality of the resulting software. Almost all of the software developed by the FMS with Laser project is in assembly language rather than more sophisticated high-level languages and is not user friendly. The system has no geometric models for producing CNC programs; indeed, the programs needed to operate the highly sophisticated CNC machine tools that are the heart of any FMS system are written manually and often with very primitive languages. It is possible that this process was chosen deliberately in order to develop a flexible machine first and to consider software later. However, many biases and subtle inefficiencies can creep into a hardware system when software development is delayed.

The FMS with Laser represents an extreme case, and many Japanese software projects are both better funded and more sophisticated. Toyoda Machine Works, for example, is Japan's second largest machine tool maker; it uses geometric modeling for numerically-controlled (NC) program development, but only for turned parts. It plans to extend this capability to milled parts in the future. A somewhat more advanced system at Yamazaki Machinery Works uses a software development environment that allows an operator to input the geometry of the part. The system then puts up a graphics display of the part, prompts the operator through a series of menus, and then simulates the cutting sequence, showing cutter pathways. Finally, the system sends the program to a milling tool to make the part. While comparisons to U.S. systems are difficult, the IBM CAPOSS and GE GE2000 systems seem to share most performance characteristics.

Japan is also active in the Computer Aided Manufacturing-International (CAM-I) Robot Planner project. This program intends to develop an off-line programming system to define robot motions for multiple robots and environments, and will include the possibility of moving objects. This effort has ambitious goals, but its million-dollar annual budget translates to about ten person-years of programming effort, probably far too little to accomplish anything significant by itself. Nonetheless, it is still possible that participating companies will contribute funds on their own, which could create the required critical mass.

In contrast to their extremely limited role in mechatronics hardware research, the Japanese universities have been responsible for some of the most advanced software work in the field to date. In particular, the GEOMAP CAD modeling system developed at Tokyo University has the ability to automatically generate CNC data from model parts for operations such as hole drilling, face milling, and contour milling. Its geometric modeling program, which consists of about 200,000 lines of FORTRAN code, is similar to the PADL2 system developed at the University of Rochester. There is no indication, however, that GEOMAP has been used in industrial applications.

SUMMARY

Japanese research in mechatronics is summarized in Figure 6–1. As the figure indicates, Japan is even with or ahead of the United States in every category except software, and the software lag may be disappearing in the near future. These are not surprising conclusions, given that mechatronics is a Japanese concept. Nevertheless, some of the reasons for differences in research on components of mechatronics bear repeating.

The Japanese industry benefits from the vertical integration which enables electronic, machine tool, and automobile companies to design their own mechatronics systems. The close ties between the

Figure 6-1. Overview of Japanese Mechatronics.

CATEGORY	BASIC RESEARCH	ADVANCED DEVELOPMENT	PRODUCT IMPLEMENTATION
FMS	0 ↑	0 ↑	+ ↑
VISION	− ↑	+ ↑	+ ↑
NON-VISION	0 ↑	0 ↑	0 ↑
ASSEMBLY	0 ↑	+ ↗	+ ↗
INTELLIGENT MECHANISMS	0 ↑	+ ↑	0 →
SOFTWARE	− ↗	− ↑	− ↗
STANDARDS	0 ↑	0 ↑	N/A
MANIPULATORS	0 ↑	+ ↑	+ ↗
PRECISION MECHANISM	0 ↑	+ ↑	+ ↗

CODING SYSTEM — JAPAN COMPARED TO USA:

PRESENT STATUS

+ AHEAD
0 EVEN
− BEHIND

RATE OF CHANGE

↗ GAINING GROUND
→ HOLDING CONSTANT
↘ LOSING GROUND

designers of the product and the designers of the machinery to manufacture the product result in greater efficiency than appears to be
possible when relying exclusively on external suppliers. In addition,
the number of Japanese engineers at several different levels involved
in designing and using mechatronics systems results in more fine
tuning of products and processes than is possible in the American
manufacturing environment. Finally, the large Japanese companies
are better able to afford mechatronics-related research expenditures
than some of the small start-up American firms responsible for components of mechatronics systems.

U.S. research in specific technologies related to mechatronics has
surpassed Japanese efforts. Certain types of vision sensors and navigational devices, such as gyroscopes, are two examples. However,
U.S. work in these areas has had either a military or a very specialized orientation with little attention to broader manufacturing
applications.

Universities in both countries have difficulty organizing the type
of interdisciplinary research teams needed to explore mechatronics.
This difficulty is of less consequence to the Japanese industry, however, which supports its own research and less dependent on university research than the U.S. industry. In the industrial setting, Japanese success in interdisciplinary research and development has been
clearly documented.

7 TELECOMMUNICATIONS

The globalization of the telecommunications equipment industry is a relatively recent phenomenon. AT&T has historically controlled a dominant share of the U.S. market for telecommunications equipment, although strong competitors have begun to emerge in the past several years. Likewise, International Telephone and Telegraph (ITT) has dominated foreign equipment markets since it split off from AT&T in 1925. In the early 1980s major telecommunication equipment suppliers began to cross national borders in search of bigger revenues and greater economies of scale. AT&T established a sizeable international division; ITT made an important entry into the American market, and companies such as Siemens of West Germany and L. M. Ericcson of Sweden have been selling aggressively overseas.

Japan has the potential to be a major player in this emerging global market. NEC has sold several central office switching systems to the U.S. regional telecommunications companies and has also gained entry into the U.S. end-user telecommunications equipment market. At the same time, the Japanese took steps to introduce market competition in 1985, with the reorganization of NTT.

Exactly what role Japan will play in the telecommunications industry is still uncertain, since the transformation of NTT in Japan and the divestiture of AT&T in the United States are both recent events. One key to understanding the future of the industry is to

107

examine the quality and scope of Japanese telecommunications technology research. We did not attempt to cover the entire spectrum of telecommunications technology, but focused on those areas most likely to affect the future of United States-Japan trade competition. Thus, the topics covered in this chapter include fiber optic and digital microwave links, but not analog microwave links; digital voice modems, but not facsimile modems; and mobile radio, but not traditional broadcast radio or television. Because of the earlier report on computer science, this panel focused on Japanese hardware-related research and development and examined only that software required to make the hardware operational.

JAPANESE ORGANIZATIONS INVOLVED IN TELECOMMUNICATIONS RESEARCH

MITI, NTT, and Private Sector Telecommunications Research

The Japanese telecommunications industry consists of two segments: a domestic service industry, long dominated by NTT and now becoming more competitive; and an equipment sector, dominated by the major NTT supplier firms: Oki, Hitachi, Fujitsu, and NEC. Although NTT is now a semipublic company with some private shareholders, it is still subject to heavy regulation by the Ministry of Posts and Telecommunication (MPT). The JTECH panel examined only briefly the activities of Kokusai Telegraph and Telephone (KDD), Japan's supplier of international telecommunications services.

The domestic monopoly on telecommunications services maintained by NTT has enabled the company to charge higher rates for domestic services than the market would otherwise bear and use the surplus to subsidize R&D in the hardware sector. The Electrical Communications Laboratory (ECL), discussed in Chapter 4, is a major driving force in Japanese telecommunications technology. The ECL was recently reorganized into nine divisions, consisting of the Telecommunications Network Laboratory, the Radio Communication Network Laboratory, the Communication and Information Processing Laboratory, the Integrated Communication Laboratory, the Research Laboratory, the Electronics Laboratory, the Software Production Techniques Laboratory, the Ibaraki Laboratory (for research on components and materials), and the Atsugi Laboratory (for research on large scale integration). This reorganization reflects ECL's new priorities: integrated networks (that is, those combining voice, data and video), software, and value-added networks. Above all, the organization suggests a Japanese attempt to deal with a traditional shortcoming of the field, namely an inability to deal with entire telecommunications systems, as distinguished from individual components, subassemblies, and equipment. Telecommunications research at ECL tends to be more applied than the research even in the advanced development division of AT&T Bell Labs. ECL deemphasizes basic research in favor of innovation at levels nearer to delivered products and systems.

NTT is also launching a major telecommunications research initiative in digital networks, known as the Information Network Service Initiative (INS). The purpose is to design a system which will provide a range of innovative services, including simultaneous voice and data transmission; videotext; two-way cable television; image transfer; data bases; and advanced telephone features, such as voice storage and retrieval, touch-tone data entry, call recording, and screening. NTT has worked closely with its suppliers to develop the switching technology, the optoelectronic technologies, and the system standards underlying INS. This is one of the most significant research programs in Japanese industrial history, with a budget of approximately $120 million over the next decade.

MITI also plays an important, if sometimes overlapping, role in funding telecommunications-related research. A large percentage of the AIST budget is allocated to projects in information technology that have a direct or indirect bearing on the development of telecommunications hardware and software. Some of these projects, such as the OJRL, have been discussed in earlier chapters. In addition to AIST funding, MITI has been able to gain federal funding to assist Japanese database vendors. Foreign vendors control about 75 percent of the $500 million database services market in Japan, the highest foreign penetration of any market segment. MITI assistance to Japanese vendors has taken the form of loans at preferential rates, loan guarantees, and promotion of legislation to grant Japanese data base firms tax savings to be set aside as a reserve for future software development expenses.

In 1985 MITI introduced the Information Industry Law, which provides for specific incentives to enhance the development of the information technology industry. One section of the proposed legislation called for MITI to oversee and develop new standards and protocols to ease communications between different types of information processing equipment. MITI also plans to develop an "Interoperable Data Base System," which would make it possible for different data-processing systems to communicate with each other. The system would enable universal transfers of data, images, and text. It represents MITI encroachment into the business of setting standards for computer communications, a function which has traditionally been the responsibility of the MPT.

Finally, MITI and the MPT joined forces in 1985 to establish the Key Technology Promotion Center for research in telecommunications and information technology. This center provides equity for research and development activities and loans to Japanese telecommunications companies as well as the basic infrastructure for disseminating scientific and technical literature. Its funding comes from the dividends on government shares of NTT stock, the Japan Development Bank, other government financial institutions, and private firms.

University Research in Telecommunications

Almost all government-supported national universities have programs in communications, at least at the undergraduate level, and several at the graduate level. Overall, Japanese universities do not seem to play the leading role in driving telecommunications innovation that some American universities have played. Whereas some of the Japanese universities have made important contributions in specific base technologies, such as semiconductors and radio propagation, this is not the case in network technology. There are no Japanese counterparts to Stanford's influence on Xerox, Palo Alto Research Center's research on workstations and server-based local area networks, or to the influence of MIT and UCLA on ARPAnet.

Japanese government support for university telecommunications research lacks the coherence of comparable U.S. funding. U.S. university funding programs, such as the Stanford and MIT examples just cited, all had some sort of theme behind them, as a result of DoD's mission orientation. As the mission orientation of federal support of American universities in the network area has become more diffuse, large companies are beginning to play more of a role in defining and supporting university telecommunications research.

Japanese universities do contribute to the field, albeit in a different capacity. University professors regularly attend the MPT Telecommunications Technology Councils as well as other private advisory groups from which the ministry obtains inputs on technological developments. There appears to be an especially strong informal tie between academic and industrial research communities in research related to mobile radio technology. This is in part because Japanese university research in this area is considerably more applied than in the United States and is also closely related to NTT research programs.

EVALUATION OF JAPANESE TELECOMMUNICATIONS RESEARCH

We examine four areas of Japanese research: networks, network subsystems, mobile radio systems, and components.

Networks

Distinctions can be made among networks on the basis of several characteristics: level of intelligence, geographic scope, type of ownership, and so forth. There are essentially three types of network systems: wide area networks (WAN) and metropolitan area networks (MANs); local area networks (LANs), and value-added networks (VANs). WANs span national or international distances, while MANs span a few miles. These networks extend beyond the jurisdiction and ownership of any individual entity and are often called public networks. A LAN can extend over many miles; its determining characteristic is that it is privately owned by a single entity and all of the devices are directly connected to that network. VANs add complex functions to the basic transmission and switching capabilities of a wide or metropolitan area network. Such networks are usually public and highly computerized.

Telecommunications networks typically remain in place for more than two decades; technological changes are therefore gradual and incremental. Hence, it seems certain that most of the major network technologies of the 1990s are already recognized: cable, satellite, microwave, and optical fiber for physical links; silicon and gallium arsenide for switching.

Nevertheless, the next decade promises to be one of unusually rapid growth of the telecom industry, for several reasons. First, the deregulation of both the American and Japanese telecommunications markets has created numerous commercial opportunities for the establishment of alternative networks to compete with the old monopoly networks. Second, the number of end users is rising rapidly, particularly among business users in Japan, where the Kanji character problem is finally reaching resolution. Third, broad new applications are arising within the telecom arena, including videophone, teleconferencing, voice mail, store and forward, and facsimile and videotex. Some of these will undoubtedly succeed and generate massive telecom capacity requirements.

In assessing the Japanese effort in network system capabilities, it is helpful to refer to the International Standards Organization's seven-layer model of telecom services. Labeled the Open Systems Interconnection model (OSI), it outlines standards for each of the major services required before two users can communicate effectively over a network. The lowest layer, layer 1, is the physical link—the transmission medium and switching system between two points. Layer 2, the data link, is the ability to send an electronic pulse successfully from one end to another. Layer 3, the network layer, routes a stream of messages along a concatenation of links, sending each to its proper destination. These three layers collectively provide basic services and are usually hardware-based.

The upper four layers, by comparison, provide enhanced services and are usually software-dependent. Layer 4, the transport layer, ensures the integrity of messages sent along the path; then layer 5, the session layer sets up and manages a two-way dialogue between users on the path provided. Layer 6, called the presentation layer, provides for necessary conversions of protocols, alphabets, screen formats, etc., so that the output device at each end presents an intelligible message to a human viewer. Finally, the application layer (layer 7) enables specific software programs to be used across a network.

In the Japanese system, the Class I companies under the Telecommunications Business Law are the only ones that may own the communications lines and switches that constitute the first layer of the OSI model; these companies also provide basic services at layers 2 and 3, and may provide some enhanced services at higher levels. The Class II companies may provide basic services at layers 2 and 3, but their primary purpose is to provide enhanced services, particularly at levels 4 through 6. The uppermost layer is most often provided, if at all, by individual users and their software suppliers. The strict separation between basic and enhanced services is not found in the United States, but it is unclear whether either nation can create a comparative advantage from this difference in industry structure.

Basic Services: Layer 1. At the lowest level of the OSI model, Japan enjoys a leading position in the development of fiber optic media and component technology and has improved significantly in satellite links. These strengths reflect Japan's geography and demographics. Most of the population lives along the coasts, which favors a linear high-volume backbone fiber network, while the many islands and the susceptibility to earthquakes and typhoons suggests the use of satellite technology, which is less easily disrupted by natural disasters. Indeed, the excellent Japanese earth station technology is known and used around the world.

Japan initially attempted to exploit this earth station technology using 20 to 30 gigahertz (GHz) transmission, but propagation conditions proved insufficient to support reliable communication. After switching to the 12 to 14 GHz band and joining forces with GE, NTT encountered severe reliability problems with the non-Japanese traveling wave tubes in the satellite; consequently, despite its lead in transmission media technology, Japan is actually behind the U.S. in switching technology and in fully deployed systems.

Basic Services: Layers 2 and 3. The actual facilities used in Japan are essentially the same as those used in other countries: analog voice-grade switched and leased lines, digital leased lines at speeds up to 2.544 Mbps, switched 50 bps telex lines, X.25 packet-switched Digital Data Exchange (DDX) public data networks, and X.21 circuit-switched DDX networks. Yet Japan is fundamentally trailing the U.S. in the application of these technologies. For example, the two DDX technologies have attracted a total of only 4,500 customers, compared to 191,000 leased lines and 76,000 switched data communication subscribers.

Similarly, Common Control Interoffice Signalling (CCIS) is not yet available in Japan; this technology sends dialing information along the network separate from the content of the transmission, allowing a distinct, software-defined network to be created to perform a variety of user functions. CCIS has been available in the U.S. for years, making possible 1-800 toll-free numbers and other more advanced functions. The current Japanese research does not appear likely to enable Japan to gain much ground relative to the U.S. at these levels of the OSI model.

Enhanced Services: Value-Added Networks. The Telecommunications Business Law has opened significant opportunities for American companies to enter the Japanese market, since no limit on foreign ownership of Class II value-added network (VAN) providers is specified in the law. Still, entry into Japan will require competing against NTT, which has long offered two types of VANs. General VANs are those in which the provider may select its own set of subscriber companies; NTT currently has about ten nationwide closed subscriber group networks to serve different communities of interest, and about one hundred regional individual networks. These have typically evolved independently from one another and hence follow different protocols. NTT also provides "special" VANs, which are intended to provide limited services for an unlimited set of subscribers. Such networks exist for facsimile and telex, among other applications.

The special VANs tend to be provided by Japanese consortia, while the general VANs most often include American firms. The most notable ventures providing general VANs are the ENS and Nippon Information and Communication Corporations. The former is a joint venture between AT&T and eighteen Japanese subscriber firms, in which AT&T will provide network technology and engineering support in exchange for financing; the latter, in contrast, is a two-way venture between IBM and NTT. Other ventures are under way by teams such as IBM and Mitsubishi; GTE and Intec; Tymnet and Marubeni; and GE and NEC.

The INS is the centerpiece of Japan's current research effort in VANs. At the heart of this sprawling concept is the notion of reducing the proliferation of independent network solutions using different technologies, and replacing them with a single network fabric based entirely on digital technology for voice, data, and video transmissions. The two basic themes are to integrate the four basic classes of networks (packet-switched, circuit-switched, broadband and mobile radio) and to provide processing within the network. The current timetable for the completion of this project extends through the end of this century, although much of the major headway is likely to be made within the next decade. Several major hurdles in the underlying technology may fall within the next year or two, such as the completion of an all-digital telephone network, the development of the CS-3 advanced satellite communication system, and the completion of the F-1.6G very large optical fiber transmission system.

Many questions remain about the feasibility of INS plans, particularly the most sophisticated elements scheduled for development well into the next decade. Still, INS is without question the most ambitious integrated communications plan being attempted anywhere in the world, and it serves as a national incentive to push existing capability to new levels. What is needed now is not a series of fundamental breakthroughs in technology, but a substantial and persistent effort to exploit the current technology in commercial environments. INS represents an instance in which Japan is behind the United States in basic research and even advanced development of key technologies, but is positioned to excel in product engineering and fully deployed systems. Hence, despite its current technological deficit, Japan is poised to obtain a significant lead over the United States in network technology commercialization.

Enhanced Services: Local and Metropolitan Area Networks. Japan initially conceived of local area networks (LANs) as large manufacturing automation systems, while the U.S. approached them from the perspective of the electronic office. This difference underscores important differences in the current status of LAN and MAN technology in the two nations.

In Japan, several fiber-optic LANs operating in excess of 30 Mbps are available commercially. These networks operate three times faster than the coaxial Ethernet 10 Mbps standard for American office networks and typically carry heavy data traffic among plant floor machines in automated factories. Japan currently has little interest in developing slower networks for office-based applications similar to those prevalent in the United States, since word processing and other office-based applications still await the commercialization of Kanji-compatible systems.

In the MAN area, NTT is currently investigating fiber-optic cables as subscriber loops (i.e. trunk lines) running from central switching offices to user premises. The wavelength demultiplexing and control equipment would be located adjacent to the user premises to convert the optical signals into electronic signals and to route them to individual telephone lines. As noted in Chapter 4, this technology awaits significant improvements in integrated optics. Currently, Japan's MAN basic research parallels U.S. efforts, but its ability to convert this research into advanced development lags the expertise developed at American facilities such as Bell Labs. Still, Japan is gaining ground in the deployment of metropolitan area networks.

Network Subsystems

Network subsystems include controllers, memories, amplifiers, switches, private branch exchanges (PBXs), transmission systems, and other products that must be combined to form an intelligent network.

Switching Systems. NTT is Japan's only domestic customer for central office switching systems. It designs its own switches and relies on four other companies—NEC, Fujitsu, Hitachi, and Oki—to manufacture the switches. Concurrently, these four firms also compete in world markets, which employ standards different from those used in the Japanese networks. Hence, each firm has two separate product development efforts under way at any given time. Moreover, purchasers of central office switches require significant software and field support along with the hardware switches.

In part because their research efforts are diffused between two markets and in part because they have had difficulty providing strong field support to foreign customers, the Japanese have fared relatively poorly in exporting central office switches to the United States. Only NEC has been at all successful in placing equipment within the Bell operating companies, for example. In Europe and other world markets, where customers require basic telephone service at minimal cost, with few advanced features, the Japanese manufacturing skills have given the country an advantage over the more fully featured (but more expensive) U.S. and Canadian switches.

Japan's capability in switching systems stands to gain immensely from several of the national technology development projects. In particular, the semiconductor, software development, the FGCS, and superspeed computer system projects promise major improvements in all types of switching systems; the optoelectronics project may have some beneficial impact as well.

Transmission Systems. Because of Japan's geography and demography, firms such as NEC and Fujitsu are currently emphasizing high-capacity, long-haul lightwave systems, such as fiber optic cables. Research is focused on increasing the distances between signal repeaters along each cable, from a current 10 km gap to a proposed 100 km gap. At such distances between repeaters, network installers can lay cable once and then be much more flexible in upgrading the optic and electronic technology at the switching and repeater nodes, since such overhauls could be done in a few central locations rather than spread throughout remote networks.

The Japanese have entered the U.S. market successfully because of highly focused R&D. NEC and Fujitsu have used proven lightwave technology and aggressive customer support to win large contracts in the United States. With regard to technology, these companies have excelled at developing the high-speed multiplexers needed to combine messages from many sources onto a single long-haul trunk line operating at vastly faster speed, and then to slow and reroute them at the remote end of the line to their individual destinations.

These companies have combined their expertise with substantial marketing advances. In particular, they have actively established field support networks and American manufacturing facilities, in order to support further growth in the United States. They have been less successful with low-speed multiplexers that combine disparate messages for transmission across local networks and hence do not accelerate them to the speed required for long-distance transmission over fiber optic or digital radio media.

The national technology development projects sponsored by MITI will not have as large an effect upon transmission system technology as they will on switching system technology, since transmission systems require less electronics and intelligence to operate. Only the semiconductor and optoelectronics efforts are expected to have any significant impact. Hence, any major breakthrough would have to come directly from the transmission system suppliers, but these firms seem satisfied to use existing technology for at least two decades.

Customer Premises Equipment. Included in this product category are telephones, voice/data terminals, facsimile machines, key telephone systems (KTSs) and private branch exchanges (PBXs). The telephone segment is a mature business, in which Korea, Taiwan, and Hong Kong are the low-cost producers. Voice/data terminals, too, are a well-understood technology, but one that has not been implemented commercially. NTT is conducting field trials of these terminals through videotex and teletext systems. While videotex has been well accepted in France, it has yet to find much customer support in the U.S. or in Japan, although vendors on both shores of the Pacific continue to believe that prosperity is just around the corner. Japanese telecom manufacturers, for instance, say that deployment of videotex and teletext terminals will be in the millions by 1990, and that their cost will be reduced by more than half, to about $220 each. Japan is probably ahead of the United States in this area.

In the facsimile segment, Japanese firms are the world leaders. This preeminence derives primarily from the nation's Kanji character set, which has favored the development of such image transmission systems at the expense of electronic data transmissions. Companies such as Cannon, Sharp, NEC, Sony, and Panafax currently dominate world markets for low-cost, high-quality conventional monochromatic machines; an 8.5 × 11 image can be transmitted within one minute. These same companies are also leading the research effort to develop ultra-high-speed and color systems, and are integrating facsimile systems into INS systems as well.

Japan has had less success attempting to enter the American market for PBX systems. Unlike telephones, terminals, or faxes, private branch exchanges represent major investments by large corporate customers. They require a prolonged selling effort and are purchased only after substantial careful consideration. Three North American firms dominate the U.S. PBX market: AT&T with its Dimension line, Northern Telecom with its Meridian product, and Rolm with its CBX models. These products span a wide range in cost and functionality, from the Dimension's low-end Merlin system for small businesses to the largest CBS-II, with over 4,000 lines. NEC, Hitachi, and others are developing modular, low-cost systems, but they have had little success in the United States. One reason is that PBXs were introduced into Japan long after they were accepted in America, thereby giving the Japanese manufacturers little domestic demand to support product development.

The national semiconductor development project will have a substantial impact on all types of customer premises equipment, since all products depend on advances in semiconductor technology. Several of the other national projects will also have an effect on the PBX technology, since PBXs require significant software and computing capabilities.

Mobile Radio Systems

Mobile radios can be either cellular or noncellular. Specific products include automobile telephones, paging systems, and satellite-augmented mobile radios. The recent popularization of cellular telephone service in the United States has been spurred in part by the prominence of relatively low-cost products provided by Japanese manufacturers. Cellular telephones represent just one of many types of mobile radio systems, and the Japanese have been important movers in most. The term "mobile" refers not simply to systems in which at least one terminal is located in a moving vehicle; it also applies to hand-held portable units in which the mobility is provided by pedestrians in a variety of applications. Such applications include private networks of varying size and complexity, private-shared networks, and public networks. The basic definition of mobile service is that the communications link extends to a person at whatever location he might be, rather than to a terminal or station at a fixed site.

This definition of mobility places some important constraints on the development of appropriate technology. First, the possible range of a system, and a person's mobility within that range, are limited by the high transmission losses incurred, and the resulting low signal-to-noise ratios at which service must be provided. Second, because the network is constantly being redefined through software over a multitude of different pathways between two moving points—as opposed to a fixed line-of-sight connection between two stationary radio terminals—the delay-spread conditions that exist often produce a coherence bandwidth of less than 100 kiloherz. While this level is acceptable for voice, it is unreliable for potential high-speed data service. Third, the fast-fading characteristics induced by the motion of a receiver limits the system's ability to compensate for the impairments caused by multipath environments. Finally, mobile radio systems are quite expensive compared to similar linkups between fixed points.

Different environments for mobile communications have created different motivations for developing new product technologies. In the United States, the planning and construction of the network infrastructure is largely left to the major equipment manufacturers, such as Motorola and General Electric. The Federal Communications Commission (FCC) arbitrates and resolves conflicting requirements for the limited resource of transmission spectrum, but these activities are generally more legal than technical. Its role should not be discounted, however, since current demand far exceeds the available supply of spectrum.

The FCC's policy has major implications for the creation and institution of standards. For private or private-shared networks with no intrinsic need for interoperability, no single set of standards yet exists. In public networks, however, standards have been enacted, largely through the efforts of predivestiture AT&T. Since the breakup of the Bell System, the FCC's policy of setting standards through open dialogue with relevant companies has enabled many firms to take part in the process, each seeking to build a proprietary competitive edge into the resulting standards.

In sharp contrast to this model, the Japanese Radio Regulatory Bureau (RRB), the MPT, and private companies all play important roles in the planning and establishment of the infrastructure for mobile networks. RRB focuses on noninterconnected systems, while MPT concentrates on wireline-interconnected systems. Industrial companies participate in Telecom Technology Councils, which are considered official organs of MPT and which actively develop and propose standards. The strongest industrial factor in the planning and research of mobile radio system networks is NTT's Electrical Communications Laboratory (ECL), which has not been materially affected by the privatization of NTT or by the deregulation of the telecom market.

Mobile Radio Networks. The prominence of the government and of the government-controlled NTT in the planning and development of Japanese network infrastructures, as opposed to private sector dominance in the United States, has resulted in significant differences in the mobile radio capabilities of the two countries. The early government emphasis enabled NTT to begin commercial cellular radio telephone service in 1979, ahead of U.S. companies, but it also locked NTT into a low-capacity service that was soon surpassed by American firms. NTT's system did accommodate call transfer as a receiver moved through network cells, but it was built around a signalling system with a 300 bits per second capacity; early American systems used a 10 kilobits per second (kbps) rate when they were introduced commercially in 1983. A second factor limiting NTT's capacity was its decision to severely limit the reuse of channels after a call has left a cell. Japan appears to be holding constant in this field, neither losing nor gaining ground; its position remains slightly inferior to that of the United States.

Japan also lags, the U.S. industry in trunked radio systems, but in this area its position is deteriorating. Multichannel access systems (MCAs) were introduced simultaneously in the United States and Japan in the early 1980s, with comparable network capacities and signaling bit rates. In Japan, the RRB established an early national standard for a simple access arrangement with little room for advances in system features. In the United States, where the signaling schemes are unique to each manufacturer, flexibility is designed into systems to allow for dynamically forming and reforming subnetworks among users of a trunked system. The flexibility also allows manufacturers to upgrade the network and to add intelligence incrementally.

While the heavy government control has led to current technology deficits in Japan in cellular and trunked networks, it has also resulted in superior planning capabilities for future networks. Indeed, NTT's proposal of a 16 kbps, all-digital mobile radio telephone is more advanced than any serious plan in the United States. However, European countries are probably ahead of both the U.S. and Japan in this regard.

Mobile Radio Components.   Japan's research in digital modulation—i.e. the conversion of sound waves into electronic pulses—is advanced relative to U.S. work. Indeed, the intensity and variety of the efforts indicates that Japan is putting more distance between the two countries. The primary technique under investigation is Gaussian-Filtered Minimum Shift Keying, which offers advantages in spectral occupancy; most studies focus on 16 kbps data rates, but some exploration of rates up to 256 kbps has begun. At these rates, however, the problems caused by the multipath network severely degrade performance.

Japan holds a similar lead in the area of signal transmission and propagation, and is likewise increasing its lead. This lead stems largely from work conducted at NTT in 1967, which was later used by Bell Labs as its standard. NTT has continued to break new ground, most recently in studying the short-range propagation characteristics critical to the successful design of cell systems. Only in the use of geosynchronous satellites as repeaters in mobile networks does the United States hold a significant lead, and it is gaining ground in this area.

Japanese firms have not staked out any major leads in customer equipment, but they are comparable to the U.S. position. In both countries, the product requirements are identical: smaller and lighter units, lower power requirements, longer battery life, increased ease of use, and reduced cost. Success in meeting these requirements depends in large part on the ability of manufacturers to miniaturize the technology and pack more functions into a single integrated circuit. The total domination of Japanese firms in consumer electronics—which require similar gains—and the ongoing work of the national semiconductor development project will enable Japan to improve its position and open a lead over the United States.

Paging Systems.   NTT introduced one-way radio call service in 1968 and has been improving its technology ever since. Nevertheless, Japan continues to lag behind the United States, where firms such as Motorola have pioneered new paging technology. In the critical technology areas of binary coding and dual address, the lag was almost a decade. Currently, the United States holds a large lead in providing advanced system features and in reducing pager size and weight. Moreover, Japan is hindered by an obsolete frequency band for paging systems. It alone still uses the 250 megaherz (MHz) band, while other countries have coordinated their systems around 450 MHz and 800 MHz bands.

As is the case for the nationwide telephone network, NTT is the sole operator of Japan's paging service, and sources its equipment from a small group of outside companies—firms such as NEC, Matsushita, Oki, Mitsubishi, and Nippon Motorola. NTT has been relatively successful selling paging systems, as the number of subscribers doubled from 1979 to 1983. Because the domestic market is still much smaller and more primitive than the American market, several of the Japanese firms have begun aggressive attempts to manufacture and sell pagers for export. Their efforts have already been sufficiently successful to trigger product-dumping charges by American manufacturers.

Components

This section considers Japanese research in two component areas: semiconductors and fiber optics, the technological underpinnings of the communication and information industries. Very large-scale integration (VLSI) microelectronics and fiber optics influence, and are influenced by, the field of telecommunications in different ways. Fiber optics technology has been driven by communications applications and thus has enabled new developments in communications system. By contrast, VLSI has been both a constraining and an enabling technology. VLSI development has been driven primarily by information processing rather than communication requirements. It seems unlikely that any function unique to telecommunications will substantially influence the evolution of semiconductor integrated technology.

Semiconductor Integrated Circuits.   Many of the component functions required by future telecommunications networks will be similar to the functions that underlie information processing systems today. Assessing research on VLSI for telecommunications therefore demands examining not only circuits dedicated to communications applications, but the strength of the Japanese semiconductor IC industry in a broader context. Quantity and quality reviews of Japanese technical literature and patents reveal that the Japanese are ahead of the United States and gaining ground in both research and development in this area.

Much of the Japanese success in VLSI component technology appears to stem from a commitment to design, engineering, and manufacturing which does not exist in the same magnitude in the United States. Although design and engineering work is performed at industrial labs in both countries, in Japan the funding for corporate research continues even when near-term commercial results are not visible.

In the United States, by contrast, engineering is seen as synonymous with development, with the result that creative engineering design research depends on the prospects of at least medium-term commercial success. In the first decade of large-scale circuit integration, such an approach was viable. New ideas could be commercialized rather quickly with limited amounts of capital, and a large number of alternative approaches to problems of commercial interest were thus brought to the market. The marketplace was relied on to identify the most attractive of those alternatives. However, in the age of VLSI, the enormous requirements of capital and time needed to commercialize an idea precludes the success of such a strategy. As a result, in the United States fundamental choices as to the technological means of pursuing future component markets are often made on the basis of scientific, rather than engineering considerations. These considerations may all too often be irrelevant even to the long-term commercial viability of a component or technology.

A second consequence of the confusion between engineering and development in the United States is the decrease in the number of skilled design engineers available to pursue major research initiatives. In Japan, industrial organizations rather than universities train such engineers. The success of this approach depends on the stability of the work force. The high mobility of U.S. engineers makes it difficult if not impossible for a U.S. company to invest in extensive training solely in support of research efforts. As a result, U.S. companies must rely more heavily than their Japanese counterparts on colleges and universities for engineers skilled in design. The structure by which universities are now supported in the United States does not encourage engineering design research, the number of faculty and students skilled in successful design of VLSI components is alarmingly small, and the shortage is likely to continue.

With respect to the future, a considerable portion of the scientific and technical research community in the United States has subscribed to the idea that computational tools will eventually provide for the complete design of VLSI components. Increasingly powerful computational aids are essential to continued progress in VLSI technology, devices, and circuits. It is unlikely, however, that computational tools will ever fully supplant the designer's function at the leading edge of component technology. History already shows that more powerful computational tools serve primarily to increase the size and complexity of the components a designer can successfully implement. They do not replace the designer. Emerging Japanese ascendancy in design expertise should be of considerable concern, as design leadership is ultimately a necessary, if not sufficient, condition for commercial leadership in semiconductor companies.

Semiconductor Memory. Success in research and engineering has been reflected in Japanese success in establishing commercial preeminence in semiconductor memory over the past five years. In 1984 Japan assumed the role of market share leader for dynamic random access memory chips (DRAMs) and erasable programmable read-only memories (EPROMs). In the DRAM market, technology moves towards the highest possible component densities while maintaining the best possible device performance. However, another form of memory component, high-performance static random access memory (SRAM), is the vehicle by which semiconductor technology is advanced to its highest levels of performance. Whereas progress in silicon technology has taken prototype complementary metal oxide semiconductor (CMOS) single random access memory (SRAM) access times down to 20 nanoseconds and below, speeds an order of magnitude faster than this have been achieved in research demonstrations. Japan has long been the leader in SRAM technology and is responsible for critical achievements in three of the four basic technologies (silicon bipolar, gallium arsenide metal semiconductor field effect transistor, gallium arsenide HEMT, and silicon metal oxide semiconductor). The Japanese industry pursued a number of different paths to see which would yield the most promising results. This approach differs from that in the United States, where it appears that the choice of technology is made on the basis of device or material characteristics rather than on the basis of engineering research results.

Custom Voiceband Circuits. In addition to semiconductor memory, Japanese achievements in custom voiceband circuits and digital signal processors will figure prominently in the future of the industry. The voiceband coder/decoder (codec) serves as the interface between analog voice telephone service and the emerging digital communications environment. In the initial application of digital technology to communications, namely multiplexed pulse code modulation (PCM) transmission systems, the codec function was implemented as a high-performance analog/digital conversion system that was time-shared over a number of telephone channels. However, by the late 1970s, semiconductor technology had advanced to the stage where the codec function could be economically on a per-channel basis as a single IC chip. This allowed voice telephone communication to be treated merely as one of a wide variety of services provided by a digital telecommunications network. IC technology, together with an increased demand for data communication, had thus dramatically altered the way in which public telecommunication networks of the future were perceived.

Approximate parity exists between research efforts in Japan and the United States on custom voiceband circuits, of which codecs represent the most mature component. The market for most of these circuits is only just beginning to emerge. Whether or not the Japanese will ultimately establish a preeminent position in these components depends on whether or not the United States maintains a research commitment to the area and finds a way to manufacture these components efficiently.

Digital Signal Processing.   Digital signal processing is expected to assume a pervasive role in future information processing and communication systems. Its principal applications occur in voiceband telecommunications systems such as modems, line equalizers, and subscriber line cards, as well as in speech and image processing systems. Voiceband telecommunication functions currently represent 75 percent of the applications in what is still a small commercial market; however, speech processing is projected to reach 23 percent of the total market within five years, and image processing, 11 percent. Over that same five-year period, the market itself is projected to grow by at least a factor of four.

Overall parity exists between Japan and the United States in digital signal processing. However, the field is still in the formative stages. Whereas the United States has established a leading position in the initial commercial market for general purpose signal processing components, Japanese corporations appear to be increasingly more effective than their U.S. counterparts in defining future signal processing applications. This effectiveness seems to stem from a combination of vertical integration and a focus on, as well as an understanding of, the consumer market. In addition, Japanese companies have been more willing than their U.S. competitors to support long-term research programs that result in the actual fabrication of signal processing components. This has enabled them to establish a lead in design expertise which may provide the basis for future commercial dominance.

Fiber Optics.   Japan has been particularly aggressive in exploiting fiber optics as a means of reducing the costs of information transmission. Three areas of Japanese fiber optics are expected to be significant in the future of the industry: optical fibers, optoelectronic devices, and transceiver electronics.

Optical Fibers. The optical transmission of information was first proposed in Japan in a patent issued in 1936. Subsequent work was sporadic until the mid-1970s. Following the invention of the Modified Chemical Vapor Deposition (MCVD) method of fiber preform formation at Bell Labs in 1974, Japanese research into means for economically fabricating optical fibers began in earnest. Initial efforts focused on the reduction of losses in MCVD fibers. Then, in 1977, a new method of fiber preform fabrication, Vapor Phase Axial Deposition (VAD), was invented at NTT. This method is the standard means by which low-loss broadband optical fibers are now manufactured in Japan. The successful development of the VAD method to its present state as a manufacturing technology was based on detailed theoretical and experimental studies and modeling of the materials and processes relevant to this method.

Japan is now addressing the short-term problems of fabrication of single-mode fibers at long optical wavelengths (1.55–1.6 microns), the reduction of costs associated with fiber production, and the improvement in optical fiber reliability.

Cost reduction is regarded as a major R&D objective. It is presumed in Japan that the large-scale use of optical fibers in public subscriber loops depends on a reduction in the cost of fiber optic cables to a level comparable to that of copper twisted pairs, and the Japanese industry is strongly committed to accomplishing that objective. Specific results expected from current research include a reduction in the cost of dopants by means of substitute materials, the high speed fabrication of large preforms, a reduction in the cost of coating materials, and the development of high-speed drawing techniques.

Long-term optical fiber research in Japan is now focused on alternatives to silica glass as the medium by which optical fibers are formed and on means by which the existing materials can be more efficiently exploited. Efforts to better the use of existing fibers include work on single-polarization, single-mode fibers, and research into techniques that exploit the nonlinear optical properties of fibers. Japanese research on optical fibers is considered comparable to U.S. activity.

Optoelectronic Devices. Optoelectronic components, such as lasers, light-emitting diodes, and photodetectors, provide the interfaces between optical and electronic media in fiber-optic telecommunication networks. The status of research in Japan on these devices was documented in Chapter 4. In those areas pertinent to telecommunications technology, Japan is ahead of the United States and gaining ground.

Transceiver Electronics. The third area of component technology critical to the pervasive use of optical fibers as a low-cost transmission medium is that of the electronic circuits used to implement the transmitting and receiving functions. The leading edge of this technology is driven by the requirements for optical transmission systems operating at the highest possible rates. The repeater electronics generally limit the achievable transmission rates.

The principal functions that must be provided by the transmit and receive electronics are low noise amplification, comparison, synchronization, and laser modulation. To date, these functions have generally been implemented with discrete components, and the most crucial and difficult to manufacture of these components are supplied largely by the Japanese semiconductor industry. A good example is the low-noise, high-frequency transistors used as front ends in circuits that amplify the low-level photodetector output.

Attempts to integrate the repeater function have been focused in two directions. In the United States, considerable effort has been devoted to optoelectronic ICs, the monolithic integration of electronic devices, and optoelectronic components. The Japanese, by contrast, have demonstrated extremely impressive results in by extending relatively conventional integrated circuit technology to the levels of performance required for broadband fiber-optic communications.

SUMMARY

Japanese research in telecommunications is summarized in Figure 7-1. Japanese performance improves as the reader moves diagonally across the chart from upper left corner (basic research on networks) to the lower right corner (applied development in components). In other words, Japanese research is stronger at lower levels of integration; although networks continue to be a problem area, Japanese performance in components has been very competitive. Work in subsystems varies, with very strong Japanese programs at the less sophisticated end, and less impressive performance in high-end products

Figure 7-1. Overview of Japanese Telecommunications Technology.

CATEGORY	BASIC RESEARCH	ADVANCED DEVELOPMENT	PRODUCT DEVELOPMENT
NETWORKS	– ↘	– ↑	– ↑
NETWORK SUBSYSTEMS	– ↑	– ↑	0↑
MOBILE RADIO SYSTEMS	+↑	0↑	0↑
COMPONENTS INTEGRATED CIRCUITS		+↑	
FIBER OPTICS		+↑	

CODING SYSTEM – JAPAN COMPARED TO USA:

PRESENT STATUS

+ AHEAD
0 EVEN
– BEHIND

RATE OF CHANGE

↗ GAINING GROUND
→ HOLDING CONSTANT
↘ LOSING GROUND

such as central office switches. As in other sectors, the quality of Japanese research is stronger in applied than in basic areas.

The reasons for Japanese success in developing and marketing telecommunications technology include the ability to draw on a strong electronics sector, the commitment to constant manufacturing improvements, and the allocation of resources to research on products for foreign as well as domestic markets.

Although Japanese university research in networks does not match the strength of major U.S. programs, in other areas, such as radio propagation, informal ties between universities and industry have clearly benefited the industry. Japanese communications professors also influence the direction of the industry through participation in national councils and advisory boards.

Those areas where Japanese performance has been less impressive include networks, which call for basic research at the systems level, and high-end subsystems, such as central office switches, which require sophisticated software. The Japanese industry recognizes both of these areas as weaknesses and has taken steps to close the gap. Software improvements expected over the next decade will benefit the telecommunications as well as other Japanese industries (see the Appendix). Likewise, the NTT-sponsored INS program, too young to evaluate, may put the Japanese in an extremely strong position in networks.

8 BIOTECHNOLOGY

The last fifteen years have given rise to an explosion in knowledge in the area of life sciences. That explosion was fueled by the development of a new battery of technologies referred to as recombinant deoxyribonucleic acid (DNA) technologies and hybridoma technologies. The application of these technologies to the development of commercially important products is referred to as the new biotechnology. Global interest in the new biotechnology has been expanding rapidly in the last several years. Thus far, only a few products have come out of the application of new biotechnology; however, virtually every large industrial company is now committing significant resources to biotechnology research and development. Estimates of the worldwide biotechnology market range from $15 to $200 billion by the year 2000. Specialty products of relatively high value are likely to appear first, with the chemicals, feedstocks, and biomass conversion taking a longer time to develop.

The role of DNA as the source of genetic material was discovered in 1944, and within nine years James Watson and Francis Crick had isolated DNA's double helix structure. Nirenberg followed in 1961 by determining the genetic code. It was not until 1970, however, that a scientific breakthrough with significant commercial implications was made, in the area of restrictive enzymes. In that year, Nathans and Smith demonstrated that these enzymes—also called restriction endonucleases—are able to cleave DNA molecules along their longitudinal axis and then to splice different molecule segments together.

A second advance with important commercial aspects was the development of a class of molecules called transferable plasmids. These molecules are small circular strands of approximately 4,000 base pairs of DNA nucleotides that reside inside a bacterial cell. These plasmids are not part of the bacterial chromosome (which may have 4 million base pairs of nucleotides) and can be replicated independently.

These two developments led directly to the first cloning of a gene in 1973. Essentially, cloning refers to the introduction of a DNA fragment containing a gene into a transferable plasmid, which allows scientists to replicate the gene under controlled conditions. The key to this procedure is the use of the same restriction endonuclease for both the plasmid and the target DNA fragment containing the gene to be cloned. This assures that the ends of the DNA fragment will be complementary to the ends of the cleaved plasmid DNA, so that these strands can be joined together. The resulting hybrid DNA plasmid can then be transferred into a bacterial cell where it will direct the production of large quantities of the desired protein.

Cloning, by changing the information in the genes which direct the formation of proteins, can be used to alter the structure of proteins (molecular enzymes). Such protein engineering can be used for virtually any enzyme by first cloning the gene for the enzyme and then applying the appropriate gene splicing and engineering procedures to interject new genetic information.

The promise of important commercial applications became evident when, within only a few years after the first gene was cloned, over 200 start-up companies were established in the United States to exploit the nascent genetic engineering market. Important legal decisions helped pave the way for these start-ups to establish profitable markets. In 1980, the Supreme Court gave early entrants a competitive edge by ruling that Chakrabarty (and by extension, other companies) could patent a microorganism. Cohen/Boyer expanded this protection to processes by patenting aspects of recombinant DNA technology. By 1981, the first commercial products of such technology were introduced; these included insulin, interferon, and animal vaccines. The market has continued to grow rapidly; in 1983, U.S. private sector investments in the field first exceeded $1 billion.

Although critical advances originated in the United States, biotechnology quickly became a global industry. This was primarily due to the low transportation costs of finished goods, the high capital requirements needed to support large-scale research projects, and the similarity of customer requirements around the world.

The influx of venture capital funds in the United States in the last half of the 1970s and the first half of the 1980s enabled American firms to stake out early leads in many aspects of biotechnology. Because their financial and managerial resources have been limited, however, they have not been able to lock out foreign competition.

The Japanese recognized the potential importance of genetic engineering early on. Even before the first successful cloning of a gene, Japan's Science and Technology Council advised the government that life sciences were worthy of special emphasis. Little action was taken, however, until the United States granted the Cohen/Boyer patent for developing recombinant DNA vectors. Japan feared that this patent would act as a barrier to its future entry into the U.S. market. At that point, the Japanese government officially targeted biotechnology as an area of special emphasis, and five leading Japanese chemical companies created the Biotechnology Forum.

The biotechnology industry has evolved differently in Japan than in the United States. The energetic beginning of biotechnology in the United States resulted from the flood of venture capital into small, start-up companies. Since Japan has no venture capital markets, the primary actors in Japanese biotechnology have been the large, established chemical, pharmaceutical, and food companies. These companies bring to the industry certain comparative advantages not necessarily shared by their American counterparts. For example, they have extensive experience in a food industry which is based on fermented products such as soy sauce, tofu, sake, and the like. The food companies that have developed expertise in fermentation—which is simply the ability to change substances through the use of microorganisms—clearly have an interest in developing more modern approaches toward their basic manufacturing processes. For example, the Kikkoman Corporation has developed an experimental bioreactor that allows it to ferment soy sauce from soybeans and other ingredients within one week, instead of the six to eight months required by older methods. More generally, Japanese fermentation industries already provide commercial sources of enzymes, synthetic amino acids, and antibiotics.

At the same time, large Japanese companies did face certain start-up problems. One of these was the shortage of trained researchers. When Suntory, the Japanese brewery, decided to start a biotechnology research program, it had to recruit twenty Japanese researchers then working in other countries. In 1984, fully one-third of its researchers were women—an extraordinarily high proportion for Japan—indicating the difficulty the company has had in finding qualified personnel.

Because the industry's economic structure has favored companies with a global perspective, both the Japanese and American entrants have actively pursued international joint ventures. For example, Shionogi & Co. used recombinant DNA techniques in testing its human insulin product; the technology was licensed from two American firms, Eli Lilly & Co. and Genentech. In another joint venture the liquor company was the world's first company to use a synthetic gene to produce gamma interferon, an anti-cancer agent. Suntory has signed an agreement with Schering-Plough Corporation, an American pharmaceutical company. Schering will gain access to Suntory's interferon technology, and Suntory will receive technology and assistance in running clinical tests, an area in which Schering-Plough has years of experience and Suntory almost none.

The degree to which the Japanese biotechnology industry has benefited from joint ventures with foreign partners is reflected in Japanese patent applications. Foreigners filed for only 19 percent of all Japanese patents from 1973 to 1981; in biotechnology, however, they filed for 76 percent of Japanese patents during the same period, indicating the importance of imported technology during that period. The Japanese industry has since learned from its imports and has developed an indigenous R&D capability; the amount of foreign research has declined accordingly. The patent statistics show that between 1981 and 1983, foreign applications had dropped to 55 percent of biotechnology patents, lower than the earlier figure of 76 percent, but still higher than the average across all fields.

JAPANESE ORGANIZATIONS INVOLVED IN BIOTECHNOLOGY R&D

MITI Support and Industry-Sponsored Biotechnology Research

AIST targeted biotechnology as an area for research emphasis in 1981, with the announcement of the Research and Development

Project for Basic Technology in Future Industries. This project included a blueprint for a national program in biotechnology. The development of three technologies was specified for research in the ensuing decade: bioreactors, large-scale cell cultivation, and recombinant DNA. AIST established the Bioindustry Office to coordinate the research programs it sponsors at private companies in these three areas. In 1984, CST expanded on Japan's goals for biotechnology research in the following decade. The Council emphasized gene regulation, analysis and synthesis of DNA, design and modification of proteins, chromosome engineering, cell and organelle modification, and alteration of components of cells and tissues.

MITI has budgeted about $130 million to the ten-year national research project and also conducts its own research at the Fermentation Research Institute, which was established in 1940. Other sources—such as the Ministry of Education, Science and Culture—also sponsor biotechnology research, bringing the total research budget much higher.

By 1982 total Japanese annual spending on biotechnology was close to $40 million. By comparison, that same year France, Germany and Great Britain each spent over $50 million, and the United States spent $511 million. Important distinctions existed between the United States and Japan not only in aggregate funding levels, but also in the allocation of funds. The Japanese government awards funds to corporations, who use the money to develop proprietary products and processes in the field. The U.S. government allocated only $6.4 million of total U.S. biotechnology funding toward this purpose. The remaining half-billion dollars went to basic research in universities. Hence, the American government funding emphasizes the development and understanding of basic biotechnology principles, while the Japanese focus is on pragmatic, commercially promising applied research.

Trade Associations

Several large trade organizations coordinate Japanese private sector biotechnology research. One is the Research Association for Biotechnology (RAB), which was established in 1981 by fourteen chemical companies. The association's stated purpose is to conduct research in the three areas targeted by MITI through collaborative efforts among the member companies and academic institutions. Research is conducted both at the member companies' labs and in university environments. Members of RAB's bioreactors group include Mitsubishi Chemical, Kao Soap, Mitsui Petrochemical, and Daical Chemical; members in the large-scale cell cultivation group include Kyowa Hakko Kogyo, Takeda Chemical, and Toyo Jozo; and members of the rDNA group include Sumitomo Chemical and the Mitsubishi-Kasei Institute of Life Sciences.

A second important private sector organization is the Japanese Association of Industrial Fermentation (JAIF), which dates to the onset of World War II. It has over 300 sustaining company members and over 10,000 individual members. In 1983, JAIF formed the Bioindustry Development Center (Bidec), which has approximately 120 affiliated companies. Bidec serves as a clearinghouse of information for the entire industry. Through eight operating subcommittees, it collects information, sponsors symposia, and publishes newsletters. The newsletters, published in English to promote international cooperation, contain general information on research conducted by member companies, specific news items extracted from industrial journals, and industry-related announcements.

University Research

The academic community in biotechnology consists of thirteen organizations within nine universities, all of which play an important role in training biochemical engineers. Approximately 80 full-time faculty (including 70 at the Ph.D. level), 67 professional staff, 55 Ph.D candidates, and 295 master's or bachelor's degree candidates were conducting research within these programs in 1984. By comparison, American universities had 79 faculty, of which only 28 had doctorates; the remainder of the personnel were professional staff, Ph.D. candidates, and master's or bachelor's degree candidates.

The unusually low proportion of Ph.D. to master's candidates in Japan illustrates the fairly low priority Japan places on basic research. The Japanese market demand for Ph.D's in this field is correspondingly low; instead Japanese companies prefer to hire students at the master's level and provide specialized training relevant to proprietary research projects. This practice is possible primarily because of the custom of lifetime employment, in which professionals generally spend their entire careers with one company. From the Japanese company's perspective, training is a wise if not a necessary investment.

The largest Japanese programs are at Hiroshima, Osaka, and Nagoya universities. As in the United States, these programs are often split among several different academic departments; in the United States, however, the lead department is often chemical engineering, while in Japan it is more often agricultural chemistry, applied microbiology, or fermentation technology.

EVALUATION OF JAPANESE BIOTECHNOLOGY RESEARCH

This chapter focuses on biomedical process technology, biosensors, cell culture technology, protein engineering, and recombinant DNA. Japanese research in each of these areas is summarized below.

Biochemical Process Technology

This field involves the application of new technology to the development of products from fermentation or bioconversion. Important products include amino acids, nucleotides, antibiotics, and chemotherapeutic agents. Japan aims to become the worldwide leader in the biochemical process industry and has a comparative advantage based on experience in food processing and low labor costs. Modern fermentation technology began in the years immediately following World War II, with the development of process technology for the manufacture of amino acids and nucleotide products for the food industry. Japan was able to gain an early and sustained advantage over other nations by using inexpensive labor to mutate large numbers of organisms and then selecting those that produced the highest volume of desired molecules.

The second major stage for Japan's biotechnology industry involved the attempt to build an infrastructure for the efficient production of known antibiotics. Foreign companies established joint ventures to build production facilities in Japan, believing the country to be a major developing market that could best be served by local production. Following the success of Japan's antibiotics manufacturing strategy, Japanese companies attempted to move into the manufacturing of more commodity products, such as single-cell proteins, citric acid, monosodium glutamate, and lysine. Japan succeeded in developing a low-cost technology for making amino acids but failed in its single-cell protein attempts because the feedstock for that technology was seen as an unacceptable pollutant by communities surrounding manufacturing facilities.

By the early 1970s, labor had become much more expensive, eroding one of Japan's traditional advantages. The industry then broadened its focus, using new technology to develop new therapeutic agents based on original Japanese research and finding new methods for improving enzymes as catalysts. Some of these efforts were not commercially successful, in part because they were motivated by technological feasibility and not by market economics. For example, Japan developed a technique for isomerizing glucose into fructose in the manufacturing of high-fructose corn syrup, but was unable to commercialize the technology. U.S. companies, however, created a $2 billion business from the same process. Similarly, Japan has been unable to generate major business opportunities from cell immobilization technology it has developed. Instead, the British firm ICI, Ltd. now leads this market.

Through the mid-1980s, the Japanese mastered the low-cost commodity manufacturing processes, such as for amino acids, but still encountered significant difficulty in attempting to enter product areas requiring more specialized technology. The Japanese have recognized their inability to develop products with improved performance parameters, and have licensed such products extensively from the United States and other countries.

The Japanese have still had difficulty marketing their new products overseas. Unlike the amino acid market, which is highly price-sensitive and which has little need for service, the markets for new products in such areas as diagnostic enzymes, pharmaceuticals, and industrial enzymes require substantial distribution and service support by the manufacturer. Japanese biotechnology companies underestimated the difficulty and complexity of setting up service and support for distant markets, very much the way Japanese telecommunications companies underestimated the difficulty of selling central office switches to U.S. customers.

The Japanese biotechnology industry has attempted to overcome these problems through a variety of actions. It has built factories in foreign countries to ease the distribution and servicing difficulties and has established joint ventures with companies that already have strong marketing and service operations. More recently, Japanese industrial giants have signed research contracts with smaller entrepreneurial companies to access new technology. Whether Japan can dominate the worldwide biochemical processing business depends upon its ability to accomplish several major tasks simultaneously. It must refine its technological base in order to become a low-cost producer of newer specialty products; it must continue to distribute its manufacturing capabilities worldwide; and it must retain open channels with Western sources of technology, including small companies and major universities. By the end of 1986, Japan had made significant progress but was still behind U.S. capabilities.

BIOSENSORS

Biosensors are products used in a variety of applications to detect levels of specific chemicals in an organic system. They include potentiometric or amperometric electrodes, chemically sensitive electronic devices, enzyme thermistors, and optoelectronic devices. These are often used in medicine, for hospital and clinical laboratories as high throughput multichannel analyzers and emergency use single-function analyzers. Japanese researchers are developing at least four different types of biosensors for both biomedical and process applications. Most current research activity is directed at the biomedical sensor market, where the near-term need appears greatest. Each of these types of sensors measures a different effect of a chemical reaction that takes place upon a membrane. The four configurations are: direct electrodes, indirect electrodes, chemically sensitive transistors, enzyme thermistors, and optoelectronic devices.

Direct electrodes are either potentiometric electrodes, which measure the change in voltage, or amperometric electrodes, which measure the shift in current. The measured effect transpires directly upon the electrode of a cell. More commonly, potentiometric and amperometric devices are indirect, with the surface effect of a reaction resulting in a shift in the concentration within a cell of the chemical produced in the reaction. Indirect electrodes are often used for such substances as glucose, which can be converted enzymatically to oxygen or hydrogen peroxide; these products can be measured directly.

Chemically sensitive transistors use microelectronic devices, such as field effect transistors, to measure changes in chemical concentrations.

Enzyme thermistors measure the thermal (temperature) effect of surface reactions. So far, they have limited commercial application.

Optoelectronic sensors are typically sensitive to changes in the fluorescence or transparency of substances as a result of biological processes. Optoelectronic sensors are used to measure changes in other substances, a wholly different application from the information transmission examined in Chapter 4.

The development of medical sensors can be separated into four stages. Stage 0 consists of current analytical systems. Stage 1, where Japanese work is concentrated, consists of sensors for use with reagents in analyzers. Stage 2 consists of sensors which can be mass manufactured cheaply and reliably. Stage 3 consists of noninvasive sensing technologies, including imaging and permanent and semi-permanent in vivo sensors. These are still technologies of the future.

The Japanese government has not targeted biosensors for special attention, and financial support for this area of biotechnology is quite limited. MITI believes that the current level of industrial research is sufficient to ensure adequate progress in the field without government involvement. However, not all Japanese companies who have been successful in Stage 1 biosensors will be able to move on to more sophisticated products. One of the key skills which will be required for marketing second generation sensors is mass production. This will give the electronics companies a significant edge over the biotechnology companies; firms such as Toshiba and Hitachi already have well-respected names in the medical products area. Since the electronics companies have limited capability in biochemistry, their approach to second generation sensors is likely to concentrate on microamperometric enzyme electrodes, Immunosensor Field Effect Transmitters (ISFETs), and enzyme field effect transistors (FETs). Immunosensors based on antibody/antigen reactions will lag behind until companies gain significant experience in biochemistry. In general, the innovation in Japanese corporations on biosensors has been focused on solutions to manufacturing problems, not on generating new concepts.

The quality of university research varies considerably; the most productive academic groups are those in close contact with industry through joint research projects. The academic groups working on biosensors have good international contacts, but are small in number and narrowly focused on short-term commercially relevant projects.

Electrodes. Historically, most development work has concentrated on the indirect electrodes—potentiometric and amperometric sensors. These typically consist of an enclosed cell containing an electrolyte and a secondary detector. One end of the cell is closed by a selectively permeable membrane. Enzyme (protein) molecules on the membrane act as catalysts, responding to a surface reaction by producing a molecule—the electrolyte—that passes through the membrane. This electrolyte is detected by the sensor, which causes the cell's electrical potential to change. In theory, the degree to which the potential changes is directly related to the size of the surface reaction.

The problem with these sensors is that they are fragile, since the chemical integrity of the enzyme may deteriorate at high temperatures or at extreme pH levels. Improving the robustness of this type of sensor, therefore, depends upon stabilizing the enzyme in a broad range of conditions. This problem is more germane to process applications in industrial environments than it is to medical applications in controlled clinical environments, but other problems exist in medical environments. In particular, the indirect electrical devices are bulky and must retain their sensing capabilities over extended period of time, which involves developing enzymes that are stable through long durations.

The Japanese have been able to leverage their semiconductor expertise in developing more robust indirect electrical sensors. For example, a variety of ion-selective and semiconductor devices have been used to detect the secondary effect of the enzymatic reaction on the primary membrane. The speed and sensitivity of these devices was surprisingly good through the initial stages of development. Potential problems with these devices, however, are that they are nonlinear and frequently require recalibration.

Research has expanded in recent years on direct devices, in which one or more electrodes of a cell are modified by treating the surface with a biological material such as an antibody or an enzyme. These cells often include a "mediator" that serves as a transfer agent for electrons between the biological material and the electrode surface. Little work on these direct cells is under way in Japan; most is undertaken in Britain and in the United States. For example, the University of Pittsburgh has specialized in biomedical sensor applications such as in vivo measurement of glucose levels. These devices suffer the same instability problems as the indirect electrodes.

In order to circumvent the instability problems, some recent efforts have focused on the possible use of amperometric devices to measure electrolyte concentration through oxidation, without the need for any intermediary. In principle, these amperometric devices would offer both selectivity and stability; unfortunately, the commercialization process is difficult. Specifically, in complex biological systems, more than one electrode process—including the desired one—may contribute to an observed current, and no way now exists to select the desired current from the other possible currents, thus reducing the device's sensitivity. The complex nature of any organic system poses significant stability problems, even for common and well understood enzymes, due to parasitic electrochemical reactions that may generate spurious signals from the cell.

Japanese companies involved in developing amperometric sensors include Fuji, Hitachi, Matsushita, NEC, Omron Tateisi, Toyo Jozo, and Toshiba. Most of the work concentrates on measuring glucose, uric acid, lactic acid, and other fairly simple chemicals. This research encompasses a variety of approaches, ranging from direct sensors based on electron transfer (Matsushita) to indirect systems using hydrogen peroxide as the mediating agent (Fuji, Omron Tateisi). Matsushita's work exploits two innovations: the use of a reduction-oxidation (redox) system to transfer electrons directly from the enzyme glucose oxidase, and the application of a thin film electrode. The redox system appears to be more efficient than the conventional use of a mediating agent such as hydrogen peroxide, in which the electrons would be transferred by oxygen. The thin film electrode is not as sensitive as other electrodes, but it is inexpensive and can be adapted readily to several other oxidase enzymes.

Academic groups have also undertaken significant efforts in amperometric electrode research, often in conjunction with industrial companies. For example, the Tokyo Institute of Technology has worked with NEC to develop a micro-oxygen device using gold and silver electrodes. This work paves the way for miniaturized sensors, which can ultimately result in inexpensive multienzyme sensors. Before this development can be commercialized for in vivo use, the fundamental problems of stability and biocompatibility must be overcome.

Chemically Sensitive Transistors. The basic drawbacks to current potentiometric and amperometric devices—their instability and insensitivity—are largely avoidable through the use of other biosensor technologies. In particular, chemically sensitive devices hold several attractions to companies interested in developing and selling biosensors, especially for biomedical applications. Devices such as field effect transistors (FETs) contain extremely small sensing elements and thus have excellent prospects for overall miniaturization. Moreover, they can be used to detect and measure virtually any substance, regardless of ionization, electroactivity, or chemical reactivity. A third advantage is that these devices hold significant potential for direct integration with signal processing microelectronics.

This type of device is typically a cell, similar to that used in an indirect electrode. The cell generally consists of a selectively permeable membrane and a FET acting as a detector. Selectivity of the membranes used can be improved significantly through the attachment of ion-specific ligands such as crown ethers. Moreover, membranes with such ligands tend to be more stable than those with enzymes or other biological agents, since the nonorganic ligands are less likely to deteriorate in harsh environments or in biologically dynamic systems.

Research is under way to produce multiple sensors on a single FET chip. The sensors may all be of the same kind, which enables a device to eliminate readings from any individual sensor that has failed or become poorly calibrated, thus improving the reliability of the average result. Alternatively, placing sensors of different types onto a single chip would allow a single cell to obtain a broader wealth of data, thus minimizing the number of probes needed. Both Hitachi and Toshiba currently offer single- and multi-ion sensors commercially, but the market is currently undeveloped. The ability to place several sensors on one chip holds great promise for biomedical applications, where space is often limited, but is less relevant for bioprocess control applications.

Academic interest in chemically sensitive transistors has increased in recent years. The Tokyo Institute of Technology and Tohoku University are leading this drive; Tokyo is working with NEC on a urease initiative, while Tohoku has worked with Kuraray and Mitsubishi to develop several micro-FETs and membrane attachment methods. Consequently, the technology for mass-producing single- and multi-ion micro-Fet sensors appears to be close to fruition in Japan.

Mitsubishi has already fabricated a dual sensor in collaboration with Tohuko University. In addition, both Hitachi and Toshiba have products under development using devices other than FETs; Hitachi's is slated to be a nine-function device, while Toshiba has chosen to develop numerous single-function devices.

Enzyme Thermistors. Biosensors can register changes in temperature on the surface of a membrane by exhibiting altered resistance at different temperatures. When a thermistor cell with an appropriate enzyme immobilized on the membrane is exposed to a substrate, the enzyme-catalyzed reaction generates heat. The cell's resistance then changes directly—but not linearly—with the change in substrate concentration caused by the reaction. Similar devices can be constructed by substituting other substances for the enzyme, such as monoclonal antibodies.

The primary advantage of an enzyme thermistor is its simplicity; it is not prone to the interferences that plague the more complex chemically sensitive sensors. Still, it is dependent upon the stability and continued catalytic activity of the enzyme, which subjects the thermistor to the same limitations for long-term or harsh environmental conditions as other devices. This limitation is particularly severe with regard to environmental conditions, since the devices can measure very small thermal effects and therefore cannot be used in situations where the temperature is very high or fluctuate over time. These limitations are so difficult to overcome that commercialization of the technology appears remote; indeed, through 1986 neither Japanese nor other scientists had begun investigating the technology in any substantive way.

Optoelectronic Devices. The emergence of fiber-optic sensor technology promises significant improvements in biosensor technology. Fiber-optic sensors are generally simple, sensitive, and robust devices compared to other technologies. Most optoelectronic devices consist of an optical fiber with a fluorescence at one end. Light traveling along the fiber excites the fluorescence either directly through a fermentation broth or indirectly via a chemical susceptor, which in turn is sensitive to changes in the fermentation broth. Sensors that use a susceptor are somewhat more delicate, since they depend on the robustness and stability of the chemical susceptor itself; still, the inherent stability of the optical fiber provides a major advantage, one that several companies and labs have begun pursuing actively. Indeed, by 1985 several prototypes were in use measuring cell fluorescence, pH, and dissolved oxygen.

The promise held by optoelectronics' exceptional stability and sensitivity has attracted much attention in Japanese academic circles, but little commercial interest. Nagoya University has immobilized cholesterol oxidase in such a system, and has found a linear relationship between light output and substrate concentrations, suggesting that a biosensor for using this enzyme may be practicable in the near future. Other work is under way at Osaka and Tsukuba Universities and the Tokyo Institute of Technology. It is unlikely that this work will compete directly with enzyme-based biosensors in the near future, however, since industrial firms seem convinced that direct electron transfer through complex devices such as amperometric electrodes represents a more promising technology than electron-to-light-to-electron transfer through simpler optoelectronic devices.

CELL CULTURE TECHNOLOGY

Animal cells can be used to produce important products, such as lymphokines, interferons, tissue plasminogen activators, and monoclonal antibodies for diagnostic use. This area is just beginning to develop commercial products. Historically, cell culture technology has centered on the industrial production of biomedical substances by transferring specific genes into microorganisms, growing large quantities of these organisms in a fermenter, and then harvesting and separating the desired product. Because these microorganisms divide every twenty minutes and require inexpensive nutrients, this process is readily adaptable to large-scale manufacturing.

More recently, companies have taken an active interest in directly producing specialty products by using cultured animal tissue in place of the microorganisms. Such specialty products include lymphokines, interferons, tissue plasminogen activators, urokinase, and monoclonal antibodies. The animal tissue divides only every fifteen to twenty hours and requires more expensive components such as growth factors and serums. Consequently, microorganism culture has a significant cost advantage over animal tissue culture, and through 1986 no animal tissue was used in any full-scale manufacturing operation.

The cost advantage of microorganism cultures is permanent; animal tissue will always be more expensive to use because of its division pace and its nutritional requirements. Still, animal tissue may be useful in applications where microorganism cultures are incapable of producing needed products. For example, protein modification by glycosylation will be important for the activity and metabolic stability of many peptides and proteins, and only animal cells can carry out this modification process.

In order to improve the process for manufacturing biological products through animal tissue cultures, two major technical challenges must be overcome. First, several laboratories worldwide are developing more efficient bioreactors and tissue culture strains, which would produce a greater quantity of a given product. The constraint in this regard is that as cell size increase, the cells become more fragile, limiting the degree of agitation possible in the manufacturing process.

Second, a gene amplification procedure can lead to increased levels of produced protein. Such a procedure has only been achieved in systems where a drug kills a host cell unless the gene is amplified, but this type of system is too inefficient to be useful in commercial environments. Several research groups in Japan are working to achieve gene amplification through indirect means without killing cells; the first company to succeed and patent its process will likely have a strong advantage.

Because of these limitations, animal cell culture technology is currently limited to university labs and industrial prototype designs. The technology is evolving rapidly, however, although it is still not clear that the clinical utility of targeted products will pan out as expected. The rapid evolution and proprietary nature of much of the early research complicate any attempt to evaluate the Japanese technology in a meaningful way, but it is still possible to assess the infrastructure Japan has set up in the animal cell culture field.

In the academic environment, Japan appears to trail the United States. The Japanese have not established as many basic research centers within major universities, and the best of the existing centers fall short of the standards set by Western counterparts. Only three Japanese universities—Osaka, Kyoto, and Nagoya—are actively conducting cell culture development research. Kyoto's program, part of its School of Agriculture, includes the use of cell fusion to improve the quality of plant proteins; the Osaka and Nagoya programs are more conventional. The fundamental discoveries in the field have been made in the United States and Western Europe, in some cases by expatriate Japanese scientists. Japan has made little attempt to induce these highly productive scientists to return to Japan.

Even though the quality of Japanese university-based research in this area system falls short of U.S. research, Japanese industrial work is at least at the level of the U.S. companies in animal tissue culture technology. The technology is supported most avidly by the larger petroleum and chemical companies in order to provide diversification from petrochemical products. These industrial firms generally have greater financial resources to undertake these development ventures than the small start-ups that characterize U.S. development. They are especially adept at buying or adapting innovative technologies developed elsewhere and using them to increase the value of existing products.

More than a dozen major Japanese companies have dabbled in cell culture research, some more seriously than others. Among the petrochemical giants, Toyo Soda Manufacturing is developing monoclonal antibodies for clinical use, Sumitomo Chemical is working toward animal cell production of interferon and tissue plaminogen activator, and Mitsui Petrochemical Industries is developing plant cell culture techniques. Mitsubishi Chemical has chosen to concentrate more on the basic research issues that Japanese universities have so far failed to address; its Mitsubishi Life Science Research Institute has several varied projects underway.

Other large companies pursuing animal cell culture technology include textile manufacturers such as Toray Industries (interferon) and Teijen (monoclonal antibodies). The Suntory brewery and two pharmaceutical concerns, Ajinamoto and Takeda, have also set up pilot projects. Finally, some small companies have achieved notable success, such as Hayashibara's ability to produce interferon from tumor cells propagated in hamsters.

PROTEIN ENGINEERING

New genetic techniques can be applied to produce new or improved proteins or enzymes. By improving the stability, activity, or turnover of enzymes through such techniques, companies can develop proprietary commercial advantages. The concept of protein engineering has often generated controversy, since it is in essence an attempt to improve and accelerate the evolutionary process by creating superior molecules through organic means. These superior molecules—typically enzymes—are needed to act as catalysts in the production of

valuable specialty biochemical products. Protein engineering therefore consists of the systematic altering of the structure of a protein
to create new, improved proteins.

Genetic engineering and pharmaceutical companies in both Japan
and the United States are particularly interested in protein engineering to improve the stability and performance of chemicals used in
current commercial manufacturing processes. The field is just emerging, however, and several major issues must be resolved before significant commercial use of protein engineering can take place. Moreover, the field's highly interdisciplinary nature implies that the
needed research may be stifled by artificial borders between academic disciplines within the university structure in Japan and the
United States.

The largest conceptual barrier to commercialization of protein
engineering remains the lack of understanding of the relationship between a protein's structure and its function. The three-dimensional
form assumed by the chain of amino acids that constitutes a protein restricts its function to that of enzyme, hormone, or another
use; unfortunately, the chains of amino acids in even the simplest
proteins are so long and complex that the basic structure of many
proteins is still undetermined. X-rays and two-dimensional area
detectors have begun to crack the structure of newer proteins, and
the increased use of computer graphics promises major improvements
in the near future.

As with animal cell culture technology, Japan had through 1986
relied primarily on Western sources of basic research in protein engineering. Japan has been particularly adept at acquiring the genetic
engineering and gene splicing technologies required, in large part due
to the American training many Japanese scientists have received. In
protein structural analysis, Japan has had less success importing technology, since it lacks both the sophisticated instrumentation and
highly trained crystallographers required by the field. Japan has
only a small pool of trained experts, and has had difficulty expanding; only 32 of about 1,000 members of the Japanese section of the
International Crystallographers Union identified protein engineering
as their primary interest, and many of these members were graduate
students.

The only two major centers for protein structural analysis in Japan
are the Institute for Protein Research in Osaka and the Professional
Sciences faculty at the University of Tokyo. The former lab consists

of eleven professionals with a $3.5 million annual budget, but it does not have any apparent industry contacts. At Tokyo, only five professionals are listed, but they do not have state-of-the-art equipment and rely on services contracted to the Brookhaven National Laboratory outside New York City.

Japan's weakness in protein structural analysis is likely to be its most severe drawback as it attempts to create a domestic protein engineering industry, but it has also had problems with acquiring the required computer graphics equipment. By and large, the Japanese equipment is about a year or two behind U.S. capabilities, and imports of U.S. models are of limited utility. The American models are difficult to obtain, since only one distributor of Evans & Sutherland Computer Corporation exists in Japan, and since the government supports the purchase of Japanese models. The required software, most often written for Digital Equipment Corporation VAX mainframes, must be converted to run on Japanese models; this further delays the Japanese effort.

RECOMBINANT DNA

Classical genetics refers to the interbreeding of strains within a species, selecting for desirable traits among the offspring. Recombinant DNA (rDNA) is a more deliberate construction of organisms with desirable traits. This is accomplished by connecting carefully selected genes to carrier molecules (transferable plasmids) in the laboratory. These coupled genes are then introduced to a host microorganism and joined with similar genes from different species.

While protein engineering techniques involve the use of a single species of animal or microorganism to produce or improve large amounts of a given protein, recombinant DNA enables scientists to join genes from several species together. This technology can potentially be used in a wide variety of applications, including:

- Production of large amounts of useful proteins, such as hormones, enzymes, and interferons;

- Diagnosis of inherited disorders and possible gene replacement therapy for patients with inherited diseases;

- Improvement of animal and plant stocks for agriculture;

- Development of vaccines; and

- Construction of modified microorganisms with higher efficiencies in certain fermentation processes.

The basic component material for recombinant DNA research is the gene, a segment of DNA base pairs that specifies the production of a specific protein. The actual production of the protein is governed by several factors that are roughly multiplicative in their effect: the number of copies of the gene in a cell; the stability of the maintenance of the gene in the cell; the regulation of the gene's activity, as determined by the structure of both the gene and the cell; secretion of protein from the cell. Each of these factors can be regulated to some extent through the use of classical genetics or rDNA techniques.

The U.S. government immediately perceived the potential for misuse in these new cloning and protein engineering techniques. Soon after cloning methods were first introduced in the United States, experts debated the potential danger that might accrue from new DNA combinations that might escape into the environment. As a result of this debate, NIH formulated a set of guidelines to govern rDNA research. Japan reacted to U.S. concern with even greater fear; it considered rDNA molecules potentially as dangerous as some radioisotopes and regulated rDNA accordingly. Moreover, as Japan's memory of Hiroshima and Nagasaki necessitated exceptionally stringent radioisotope regulation, Japanese rDNA regulation was likewise extremely stringent. As research in both the United States and Japan has proceeded without incident, and as the process is better understood, both the NIH guidelines and the Japanese regulations have been relaxed considerably.

Japan has long been involved in the specific technologies that underlie rDNA, particularly molecular biology, genetics, and microbiology. This strength has enabled Japan to become competitive quickly in the emerging rDNA field. Japan has also been historically strong in the fermentation industries, an advantage in working with some of the microorganisms typically used in current rDNA efforts.

Several important component technologies have evolved as rDNA research has progressed. In some of these, Japan is the world leader. For example, the Japanese were the first to discover plasmids, the circular DNA strands crucial to cloning technology. The cloning vectors used in both countries are essentially identical, and Japanese researchers continue to show impressive abilities in designing new vectors to achieve specific goals. The Japanese are also among the world leaders in improving the stability of the plasmids within a cell.

In other areas, Japan's preeminence is less visible, but the quality of research compares favorably with that of the United States. Both countries have made equivalent advances in regulating the gene's activity in an rDNA process. This regulation involves manipulating promotor sequences, upstream activating sequences, or enhancer sequences, depending on the organism, but it can also involve direct manipulation of the genetic code. The Jichi Medical University and the Cell Technology Center of the Osaka University medical school have published comparisons of the strength of various promotor sequences, for example. Japan's historical expertise in microbiology has particularly aided research efforts in gene regulation.

Awareness of areas where Japanese expertise lags that of the United States is increasing. In particular, Japan lags the United States in controlling the secretion of proteins from a cell. Japan's deficit in these areas does not stem from a lack of information; Japanese scientists are kept fully apprised of rDNA work worldwide by the large number of junior faculty Japanese scientists doing postdoctoral work in the United States and European laboratories, and the rapid dissemination of new information in both English- and Japanese-language journals. Instead, Japan has been limited by strict regulations and generally low levels of funding. The low budgets of most rDNA research programs are reflected in the lack of disposable glassware and plasticware (such as test tubes and petri dishes), the generally lower quality of equipment, and the extremely short supply of technicians; Japanese students and even junior faculty often assume tasks which in the United States would be done only by support or clerical staff.

SUMMARY

Japanese biotechnology research is summarized in Figure 8-1. As in other fields, the United States maintains a lead in basic research, but the Japanese industry is making rapid progress in all but one area of advanced development. Basic research at Japanese universities is, in many areas of biotechnology, comparable to U.S. research in quality, but the academic field is small and poorly funded. The most successful Japanese university programs are those with an applied orienta-

Figure 8-1. Overview of Japanese Biotechnology Research.

CATEGORY	BASIC RESEARCH	ADVANCED DEVELOPMENT
BIOCHEMICAL PROCESS TECHNOLOGY	N/A	−↗
BIO-SENSORS	−↗	0↗
LARGE SCALE TISSUE CULTURES	−↗	0↗
PROTEIN ENGINEERING	−↑	−↑
RECOMBINANT DNA	−↑	0↗

CODING SYSTEM — JAPAN COMPARED TO USA:

PRESENT STATUS

+ AHEAD
0 EVEN
− BEHIND

RATE OF CHANGE

↗ GAINING GROUND
→ HOLDING CONSTANT
↘ LOSING GROUND

tion. The one area where Japanese advanced development lags that of the United States, protein engineering, relies heavily on an academic field that simply does not exist in Japan.

The industry benefits from Japanese experience in fermentation technology as well as from a strong electronics sector and has been successful in exporting products whose critical competitive factor is price. Unlike biotechnology in the United States, which grew out of a strong basic research tradition in biochemistry departments of American universities, biotechnology in Japan emanates from research in the large, vertically integrated corporations.

The Japanese criticize their own research in recombinant DNA as less creative than U.S. work. The JTECH biotechnology panel did not unanimously agree with this assessment and noted many instances of original Japanese thinking. The Japanese discovery of control of gene expression by gene rearrangement in bacteria, yeast, and immunoglobulin is only one of many examples. The Japanese self-criticisms have some foundation in the uniformity of the educational system and in the university system, which does not allow faculty members full independence until they are forty to forty-five years old, and then only if they can secure a full professorship. It is perhaps significant that some of the more creative work identified in this area emerged from the research environment at the Virus Institute at Kyoto University (the exception rather than the rule), where the junior faculty had considerable independence.

EXECUTIVE SUMMARY, JTECH PANEL REPORT ON ADVANCED COMPUTING IN JAPAN

Marvin Denicoff,
Thinking Machines Corporation

The objective of this JTECH panel on advanced computing was to examine and evaluate progress made in Japanese advanced computing, encompassing but not limited to work on the Fifth Generation Computing Project from 1982 to the present. Beyond providing a general overview of the Japanese effort, the JTECH panel's determination was to focus attention on these particular research areas fundamental to realizing the broadest level announced goal of the Fifth Generation program—that goal stated as "providing for the conditions and information demands of the society of the 1990s"; this to be accomplished through the "utilization of more varied media, easy-to-use computers, higher software productivity, and the application of information technology to those areas in which existing technology has not been applied."

More particularly, in its first year of operation under the organizational banner of the Institute for New-Generation Computer Technology (ICOT) Fifth Generation more sharply defined its strategy. Its primary goal—to satisfy the anticipated information needs of the 1990s—would require accessing research results and conducting re-

The other authors of this report were: J. Goguen, SRI International; C. Hewitt, Massachusetts Institute of Technology; D. Mizell, University of Southern California; S. Rosenschein, SRI International; E. Schonberg, New York University; and J. Tenenbaum, Schlumberger. Information on ordering JTECH reports can be found at the end of JTECH Panelists.

search in such fields as VLSI, distributed processing, software engineering, Artificial Intelligence (AI), and parallel computation.

For many American and European scientists, ICOT's plan was interpreted as a major thrust to establish Japan as the leader in the fields of AI and parallel computing. One ICOT manager believes that the Fifth Generation Project is an endeavor to take Japan beyond its status of playing catch-up into areas never before explored anywhere in the world. For this Japanese manager and in the view of many American and Japanese managers and scientists, the original Fifth Generation intention was to make important, highly creative, basic research contributions to AI and parallelism.

A contrasting but equally valid view was presented by one group of prominent Japanese computer managers. They view Fifth Generation's real objectives as primarily product- or business-oriented—to foster commercially important technologies. Fundamental techniques of AI and advanced software developed in American research centers would enable the Japanese to move from American esoteric research whose direct aim was the creation of practical, utilitarian products. The most extreme declaration of this perception—attributing it to a societal trait—was a recent quotation from a Japanese senior executive of the giant C. Itoh Trading Company, Takesli Ogata. Ogata said, "I am sorry to say that in the leading edge of technologies, there is continuing advantage for the United States. Our social environment, the personality, and the character of our people and institutions is to constantly improve on a process, say from a 16K RAM to a 64K RAM to the 256K RAM. But to develop a whole new technology, to get on the leading edge, you have to look to American firms. It's a cultural thing."

These differences, both in the definition and perception of the underlying goals of the Fifth Generation Project over the first five years of a planned ten-year history, certainly added to the complexity of the evaluation task confronting the JTECH panel. Against which criteria and goal statements would progress and accomplishment be assessed?

Independent of the difficult issue of coming to grips with the initial or dynamically evolving and changing description of Fifth Generation's intention, the JTECH panelists unanimously concluded that the Japanese effort to date has produced no fundamental advances.

On the other hand, it has scored points in such important areas as: (1) establishing Japan as a full partner in the international community of computer scientists and scholars; (2) demonstrating aware-ness—beyond that of other countries and most particularly the United States of the appropriate literature and most recent progress in the entire field of computer science; and (3) creating AI and soft-ware products and tools of high quality, along with an organizational research and development planning mechanism that ensures the timely and efficient transfer of research results into products.

JTECH panel members were selected and assigned review responsi-bilities consistent with areas of expertise established by their inter-national reputations. Expertise was supplemented by review of per-tinent Japanese literature and by a trip to Japan involving a one- or two-week visit to relevant government agencies and leading academic and industrial research laboratories. The U.S. visitors testified to the warmth of their reception and the openness of the dialogue. This reception was in no small part attributed to the Japanese impression that the JTECH panelists were major figures in the American research community and committed to a free exchange of ideas. A mutual learning experience was indeed the orientation that set the tone for the Japan visit. The JTECH panelists were accompanied by a repre-sentative from the National Science Foundation and the Defense Advanced Research Projects Agency. The JTECH charter of report-ing on Fifth Generation progress led to a division of the total task into five technological sub-areas, each the responsibility of one of the involved U.S. scientists. These sub-areas were: artificial intelligence, man-machine interaction, software, machine architecture, and lan-guage for concurrent computation.

It should be understood that overlaps across these areas are re-flected in the individual papers. While there is general agreement on the basic findings of the panel, there are differences of opinion or a variation in perception on several scientific issues. Rather than push for consistency, we have chosen to let the reader come to conclu-sions. Our view was that the underlying technological arguments con-stitute an exciting intellectual debate.

One example of controversy had to do with the Japanese success in developing AI products. The Fifth Generation Project has been heavily involved with speech and image processing and has produced tangible results in such fields as language translation, speech and character recognition, document and mail processing, and expert systems. Furthermore, these developments frequently involve the design of special-purpose hardware to facilitate implementation.

There is no disagreement that these Fifth Generation results represent the state of the art and compare favorably with similar U.S. products, even teaching us a lesson in the speed of development and smooth industrial coupling of these commercially directed efforts. However, these products share with their American counterparts an inherent fragility. There are strong doubts as to whether these efforts are extensible or likely to be generally useful without enormous additional research and development investments. The movement of such domain-tailored products to other contexts where issues of language and image file size and complexity is an inherent property has, in the past, always constituted a scaling problem that has frustrated AI researchers. Given the limitations of the Fifth Generation short-term approach, certain members of the JTECH team believe that Japan's proven capacity for superb engineering and research focus will discover "fixes" for these concerns.

In any case, the Fifth Generation effort will have provided the Japanese with a set of skills and a knowledge that will, along with their traditional prowess at operating at the advanced development spectrum of the research and development cycle, exploit product-oriented breakthroughs whichever the country of first invention. Other members of the JTECH study group, while acknowledging the Japanese accomplishments, argue that this progress represents a diversion from the original goals of the Fifth Generation and has been bought at the expense of, and substitute for, working on the core problems of AI. In effect, this view sees Japan as having forfeited its potential role of contributing to fundamental advances and deep solutions to make short-term, ephemeral progress.

Another difference of opinion held by the JTECH panelists concerns the Japanese design of base languages for concurrent computation. Contrary to a widely held belief in the U.S. computer science community that the Fifth Generation Project is based on Prolog, the Japanese have moved away from strict logic programming to innovate a new concurrent language called Flat Guarded Horn Clauses (FGHC). FGHC provides a declarative programming style for machine interface that seems to improve on Prolog in terms of simplicity, portability, and "debug-ability." The essence of this language thrust is the Japanese commitment to develop a truly workable, efficient parallel programming structure. FGHC is intended to serve the functions of facilitating the development of higher level languages and in the design of highly concurrent architectures. FGHC, by providing a precise language interface between hardware and software, satisfies the interests of ensuring awareness on the part of both hardware designers and programmers as to what each party has to know about the conventions of each other's implementation strategy.

The FGHC advocates further argue that the language is important to such AI goals as natural language processing and represents a facile basis for communication amongst all ICOT researchers. The debate in this case is, at one level, similar to the argument about the Fifth Generation concentration on short-term products; namely, that the announced goals of major achievements in AI and intelligent machines are being subverted by obsessive attention to logic programming. Admitting some plausibility to the Fifth Generation belief that logic programming affords leverage in solving the longer-term, substantive problems of AI, the questions are directed at what some people in the computer science community regard as a low probability of success and a high risk in focusing on FGHC.

The doubters are concerned about the recent domination of Fifth Generation by this concentration on FGHC. Logic programming now seems to be regarded as the focus, if not simply an integral part, of the research in the five ICOT research laboratories (Parallel Logic, Natural Language Processing, Knowledge Based Machines, Parallel Interface Machines, and Expert Systems).

In opposition to the believers (including several panel members) who hold that logic programming, and particularly FGHC, promises significant downstream advances in many AI areas and smart machine applications are the doubters. A few JTECH panelists suggest that the promise of FGHC is open to serious question and a focus on this approach has resulted in a neglect of other substantive approaches not related to FGHC. If Fifth Generation is betting a substantial portion of its budget on logic programming, the anxiety is that failure of FGHC will symbolize a failure of the early goals and the very promise of the Fifth Generation Project.

All of the panelists grant that the Japanese are performing at the level of, or ahead of, the United States in the theory and practice of logic programming. The questioning has to do with the importance or value of this achievement to higher-level objectives of the international AI community and Fifth Generation itself.

As the JTECH study identified and explored several controversial factors related to progress of the Japanese Fifth Generation Project, panelists found total agreement in being positively affected by particular real accomplishments. Japanese skill in rapidly producing utilitarian image and language understanding products, while sharing with similar U.S. inventions the previously described characteristics of fragility and unextensibility, was nonetheless impressive. The superb software engineering design work of FGHC was uniformly applauded by all panelists despite their differences on the ultimate value of logic programming to the deeper interests of AI and parallel computation. The Japanese work in conventional architectures, notably supercomputers (not an inherent part of Fifth Generation, but significantly affecting its development), continues to be world-class. The speed and high reliability of supercomputers produced by Hitachi, NEC, and Fujitsu were demonstrated to American visitors. These computers proved as exciting as their reputations.

Equally impressive was the excellence of Japanese software development in support of its supercomputers. The JTECH panelists found this software work competitive with, if not superior to, the best quality output in the United States. While at least keeping pace with America in the development of conventional supercomputers, the Japanese are clearly behind the United States in the design of large-scale multiprocessors or parallel machines. One of the panelists even noted an expression of skepticism on the part of some Japanese designers on the utility of parallelism for general-purpose computing, but all panelists paid tribute to Japanese awareness at the deepest technical levels of American progress in parallel computing. This knowledge, added to Japan's long-demonstrated capacity for supremacy in physical device/component technology, sends a message that complacency about American leadership in this area is a dangerous attitude.

Along with recognition of the described Fifth Generation Program's progress, JTECH panelists found some research gaps. Little attention, for example, is being paid to developing formal methods for proving the correctness of computer programs. Outside of Japanese efforts in Prolog, JTECH visitors saw general inattention to transformational programming. Compared to American AI laboratories, the Fifth Generation Project seems almost to have abandoned substantive research in such fundamental areas as knowledge representation and reasoning, machine learning, fine-grain parallel architectures, connectionism, and neural modeling.

The search for explanations of Fifth Generation gaps or differences between Japanese and American research activity leads to many conclusions. Takesli Ogata's statement (quoted above) which seems to attribute the absence of more fundamental computational research in Japan to endemic sociological and personality traits. JTECH panelists suggest other possible rationales. Given the significant advantage of large American R&D budgets over those of the Japanese, Japan might be making a conscious economic judgment to concentrate on projects with high, near-term payoff. Complimenting this viewpoint is the commitment by Japan, with human resources much more limited than the USA's, to assign its research personnel to the most commercially promising research areas.

Another possible explanation for the gaps is based on the Japanese conviction that the country has built a technological and organizational infrastructure that ensures a capability for rapidly assimilating research results developed in other countries. None of the JTECH panelists subscribe to the notion that the Japanese people are inherently imitators or incapable of the highest levels of creativity. If there is a Japanese sociological and psychological trait recognized by the panelists as perhaps underscoring the acknowledged commercial success in Japanese computing, it is an openness about learning from others, along with a diligence and commitment to achieving an awareness of research literature from the rest of the world, a determination to blend this borrowed knowledge with natural Japanese ingenuity, and the will to become the world's best at producing useful products.

With time, the Japanese may use this ingenuity to make their share of contributions to the world's storehouse of basic research results.

JTECH PANELISTS

COMPUTER SCIENCE (December 1984)

David H. Brandin, Chairman
Jon L. Bentley, Carnegie-Mellon University
Thomas F. Gannon, Digital Equipment Corporation
Michael A. Harrison, University of California at Berkeley
John P. Riganati, National Bureau of Standards
Frederic N. Ris, IBM Thomas J. Watson Research Center
Norman K. Sondheimer, University of Southern California
  Information Science Institute

OPTO- AND MICROELECTRONICS (May 1985)

Harry H. Wieder (Co-Chairman), University of California
  at San Diego
William E. Spicer (Co-Chairman), Stanford University
Robert S. Bauer, Xerox Palo Alto Research Center
Federico Capasso, AT&T Bell Laboratories
Douglas M. Collins, Hewlett Packard
Karl Hess, University of Illinois
Harry Kroger, Microelectronics and Computer Corporation
Robert I. Scace, National Bureau of Standards
Won-Tien Tsang, AT&T Bell Laboratories
Jerry M. Woodall, IBM Thomas J. Watson Research Center

163

ADVANCED MATERIALS (May 1986)

James Economy (Chairman), IBM
Michael Jaffe, Celanese Research Company
William J. Koros, University of Texas at Austin
Ralph M. Ottenbrite, Virginia Commonwealth University
Elsa Reichmanis, AT&T Bell Laboratories
John R. Schaefgen, DuPont Textile Fibers Pioneering Research
  Laboratory (retired)

MECHANICS (March 1985)

James L. Nevins (Chairman), Charles Stark Draper Laboratory
James S. Albus, National Bureau of Standards
Thomas O. Binford, Stanford University
J. Michael Brady, Massachusetts Institute of Technology
Michael Kutcher, IBM
P. J. MacVicar-Whelan, Boeing AI Center
G. Laurie Miller, AT&T Bell Laboratories
Lothar Rossol, GMF Robotics
Karl B. Schultz, Cincinnati-Milacron

TELECOMMUNICATIONS (May 1986)

George L. Turin (Chairman), University of California at Berkeley
William H. Davison, University of Southern California
Paul E. Green, IBM Research Center
James Mikulski, Motorola
Albert Spencer, AT&T Bell Laboratories
Bruce A. Wooley, Stanford University

BIOTECHNOLOGY (June 1985)

Dale L. Oxender (Chairman), University of Michigan
Charles L. Cooney, Massachusetts Institute of Technology
David A. Jackson, Genex Corporation
Gordon Sato, W. Alton Jones Cell Sciences Center
Reed B. Wickner, National Institute of Health
John R. Wilson, Lord Corporation

JTECH reports can be purchased either from the National Technical Information Service, 5285 Port Royal Road, Springfield, Virginia 22161 (703) 487-4650, or directly from SAIC, 1710 Goodridge Drive, P.O. Box 1303, McLean, Virginia 22012 Attn: Wendy Frieman.

INDEX

ABOUT THE AUTHORS

George Gamota, founder of the Japanese Technology (JTECH) project, is president of Thermo Electron Technologies Corporation, a senior corporate consultant for a number of companies, a member of several corporate boards of directors, and an adviser to governmental agencies. He is also a member of NASA's technical advisory committee, a fellow of the American Association for the Advancement of Science and the American Physical Society, and a senior member of the IEEE. During some twenty years of experience in managing technology, Dr. Gamota has been a professor of physics and director of the Institute of Science and Technology at the University of Michigan, director for defense research in the Department of Defense, and a member of the technical staff of Bell Laboratories. He has written extensively on technology and U.S. competitiveness. Dr. Gamota holds a Ph.D. in physics from the University of Michigan.

Wendy Frieman is director of the Asia Technology Program at Science Applications International Corporation (SAIC) in McLean, Virginia, where she is responsible for research on science, technology, and industry in East Asia. Among the program's clients are the National Science Foundation, the Department of Commerce, the U.S. Congress Office of Technology Assessment, and a number of Fortune 500 corporations. In addition to directing the Japanese Technology

Evaluation Program, on which this volume is based, Ms. Frieman has also written extensively on Chinese research and development, Chinese military modernization, and Sino-Japanese technology transfer. She has traveled and conducted field work throughout Asia as part of her research activities. Prior to her position at SAIC, Ms. Frieman was a China consultant at SRI International in Menlo Park, California. Ms. Frieman received her BA and MA from Stanford University. She speaks Mandarin Chinese as well as several European languages.